Tony Barnett was Professor of Zoology at the Australian National University until his retirement. He is internationally known for his many researches on stress and exploratory behaviour and also for his insistence on logical and scientific rigour in biological debate. He has had many years' experience of science broadcasting and continues to be a regular contributor to ABC Radio's *Science Show* and *Ockham's Razor*. His most recent book, *The Science of Life* (1998), was published to acclaim.

ALSO BY S.A. BARNETT

The Human Species
'Instinct' and 'Intelligence'
The Rat: a Study in Behavior
Modern Ethology: the Science of Animal Behavior
Biology and Freedom
The Science of Life: from Cells to Survival

SCIENCE, MYTH OR MAGIC?

A Struggle for Existence

S. Anthony Barnett

ALLEN & UNWIN

First published in 2000
Allen & Unwin
9 Atchison Street, St Leonards NSW 1590 Australia
Phone: (61 2) 8425 0100
Fax: (61 2) 9906 2218
E-mail: frontdesk@allen-unwin.com.au
Web: http://www.allen-unwin.com.au

National Library of Australia
Cataloguing-in-Publication entry:

Barnett, S. A. (Samuel Anthony), 1915– .
  Science, myth or magic?: a struggle for existence.

  Bibliography.
  Includes index.
  ISBN 1 86508 122 1.

  1. Science and civilization. 2. Science indicators.
  3. Science—Social aspects. 4. Human biology. I. Title.

001.9

Set in 11/13.5 pt Janson Text by DOCUPRO, Sydney
Printed and bound by Griffin Press Pty Ltd, Adelaide

FOREWORD

TWENTY YEARS AGO I invented a human ancestor called *Homo micturans* who evolved in Australia, got tiddly, and gave rise to the apes. We put the farrago on ABC Radio and a 'fossil beercan' in the Australian Museum with much pompous signage. To our amazement, many took it seriously: we even had a complaint from a temperance society.

Tony Barnett too has given us a new species of *Homo*, of which *Homo pugnax* is the most worrying. He wants us to contemplate it, much as I asked our poor listeners to deal with *Homo micturans* back in 1979, not as an exercise in frivolity, but to discover who we really are. There are many dispiriting versions of what the human animal is made of, the nature of our biological original sin, and we are bombarded with fresh versions of our predicament in nearly every newspaper as further 'genes for . . .' this and that are announced. One could be forgiven for inferring that people are little more than genetically based computer programs designed by Beelezebub.

Genes are certainly crucial. But, as the British biologist Steven Rose has suggested, they resemble a cake recipe in which the cooking is the all-important environmental ingredient. Such a process can produce quite different results from the same genetic prescription. If there are not single genes for eye or hair colour, how can one really justify all those 'genes for . . .' religiosity, homosexuality or some other complex behaviour?

This argument has gone on for much of the century. When I first met Professor S.A. Barnett in his large study at the

Australian National University in Canberra he was immediately willing to consider some broadcasts taking to task a few of the more implacable biological determinists such as Konrad Lorenz, B.F. Skinner and Robert Ardrey. The distinguished professor seemed rather severe in 1973 (he still does in some of his radio talks) but then I noticed the name-plate on his desk. It was in some elaborate Indian script.

Could it be that S.A. Barnett enjoyed a joke? Well, of course I now know that, beneath the formal carapace, is someone who is sometimes positively pixilated. But this is not an opinion shared by those who have experienced his withering analysis, his pronouncement that your thesis is 'hokum'! Many from Richard Dawkins to Richard Wrangham, from Oxford in England to Cambridge in Massachusetts, have been startled by his scorn. Whatever his sense of fun, Tony Barnett does not pull his punches.

But it is all in a good cause. Two good causes, in fact. One is to maintain the highest scientific standards, ideas backed by good experiment. The second is to spare humanity yet more ideology in the guise of 'science'. The twentieth century was besmirched by too many villains promoting too many egregious versions of the human condition. Tony Barnett has spent his long life taking them to task.

On that first meeting, all those years ago, we agreed on a subject for broadcast and I learned in the process that my interlocutor had written his first scripted talk for the BBC about three years before I was born: in 1941. Nearly six decades later I received the manuscript for a Science Show series which has formed the basis for this book. It shows Tony Barnett at his best: erudite, combative and entertaining. With so much now being promised by human genome projects, genetic therapies and the social changes that must flow from them, it is important to have a sceptic of his experience patrol the promises on offer. Given the willingness of Australians to allow themselves to be duped by my *Homo micturans*, who knows what other wild theories may otherwise gain currency.

Robyn Williams

To Kate, always there and urging me on.

If circumstances lead me, I will find
Where truth is hid, though it were hid indeed
Within the centre.

SHAKESPEARE, *Hamlet*

CONTENTS

ILLUSTRATIONS

PREFACE: HOPE FROM REASON

What is a man
If his chief good and market of his time
Be but to sleep and feed? a beast, no more.

SHAKESPEARE, *Hamlet*

A T THE FRONTIER OF a new millennium, the struggle for human survival demands a science which can be trusted. Scientists must not only give humanity reliable knowledge of nature: they must also state clearly what may be said, scientifically, about the human species; and what may not be said.

This book developed out of a proposal for a series of radio programs on the obstacles to spreading reliable news of science and especially of human biology. From the moment I began to write, the more hazards appeared. Some are due to the accelerating increase in difficult knowledge. Others, more dangerous, are of a different sort. They include superstition and even fraud. The most insidious are in the bogus biology, widely propagated, which—in effect—represents a human being as 'a beast, no more'.

We can, however, benefit from studying these stories.

While I was writing, I found a relevant passage in T.H. Huxley's essays. In 1894 Huxley, a master of popularisation, wrote of 'that prince of lecturers, Mr. Faraday', who was consulted by a novice speaker called upon to address a 'highly select and cultivated audience' [such as the readers of this book]. The beginner asked what he might suppose his hearers to know already. Faraday emphatically replied, 'Nothing'. Huxley continues:

To my shame as a retired veteran, who has all his life profited by this great precept of lecturing strategy, I forgot all about it just when it would have been most useful. I was fatuous enough to imagine that a number of propositions, which I thought established, and which, in fact, I had advanced without challenge on former occasions, needed no repetition.
I have endeavoured to repair my error . . .

So have I.

The present book is a sequel to *The Science of Life*, which outlines modern biology. There I assume that, for the most part, the findings of biological science and the achievements of scientists are welcome. Here, instead, I face the several forms of opposition to science and also its real limitations. The first six chapters give examples of superstition, magic and charlatanry; political platforms masquerading as scientifically based programs; and much pseudobiology. They also give rational responses to them, based on authentic science, especially in relation to what they signify for everyday life. In those chapters, some sacred cows prove to have feet of clay.

The second half of the book is on the scope and limitations of scientists and science. Throughout, I give actual instances of scientific and other scholarly practice and I try to avoid vague generalisations. These chapters are reminders that Hamlet goes on to speak of our 'capability and god-like reason'. They point to ways in which science can serve both truth and humanity in our present predicaments.

This book is therefore designed to help readers to decide how they see themselves and the world, and so to contribute to the continual debate, leading to action, essential in a democracy.

ACKNOWLEDGEMENTS

PORTIONS OF THE BOOK have been read and criticized by kind friends. Among them are Hannah Andrianopulos (genetics); Linda Barnett (fine arts); Paul Crook (history); Ewen Drummond (mathematics); Richard Mark (neurobiology); Kate Munro (medicine); Donald Walker (natural history).

Special thanks are due to the Science Unit of the Australian Broadcasting Commission and to Robyn Williams in particular. Their radio programs provide a forum in which divergent views are expressed. They have allowed me to present talks and programs on the role of science, especially the human sciences, in the community. This activity, the resulting lively correspondence (not always favourable) and encouragement by my publishers, Allen & Unwin, have supported me in attempting the present book.

The principal authors and sources mentioned in the text are listed in full in the bibliography.

PART I

MAGIC AND MYTH

Modern science emerged gradually from a combination of practical operations with magical arts. Myth and the occult are still powerful, but findings from scientific enquiry can counter the dangers in magic and in unsupported beliefs.

CHAPTER 1

FASHIONS IN FAIRY TALES

> This is the excellent foppery of the world that when we are sick in fortune, . . . we make guilty of our disasters the sun, the moon and stars; as if we were villains on necessity, fools by heavenly compulsion, knaves, thieves, and treachers by spherical predominance, drunkards, liars, and adulterers by an enforced obedience of planetary influence . . .
>
> SHAKESPEARE, *King Lear*

ONCE UPON A TIME, an English medical man, William Withering, fell in love with a woman who liked painting pictures of wild flowers. So he took up botany. He then learned of an old countrywoman in Shropshire, who had a secret cure for dropsy containing more than twenty herbs. A famous philosopher, Bertrand Russell (1872–1970), has written about such folklore.

> The study of anthropology has made us vividly aware of the mass of unfounded beliefs that influence the lives of uncivilized human beings. Illness is attributed to sorcery, failure of crops to angry gods or malignant demons . . . Eclipses and comets are held to presage disaster.

I do not know whether those herbs were gathered by the light of a full moon but, obviously, the old woman's remedy was an example of magical practices and primitive superstition—the opposite of science.

Or was it? Dropsy is an old word for oedema, which can be a sign of one kind of heart disease: congestive cardiac failure. Among the magical herbs was foxglove (*Digitalis purpurea*). In 1775, from his findings in folklore, Dr Withering was able to add foxglove to the pharmacopeia. Later, a nineteenth century poet, Sarah Hoare, in verses on the pleasures of botany, wrote appropriate lines:

And DIGITALIS wisely given,
Another proof of favouring Heaven
    Will happily display,
The rapid pulse it can abate,
The hectic flush can moderate,
And blest by him, whose will is fate,
    May give a happier day.

The original observation was itself, in an acceptable sense, scientific: it was rational and derived from experience; it may have been gradually developed, by lengthy trial and error, before success. Digitalis can be paralleled by many other 'ethnic' drugs, such as cocaine (analgesia), curare (muscle relaxation) and quinine (treatment of malaria).

Today, more such drugs are being searched for and found. Correspondingly, the disrespect shown by Europeans toward tribal people (called 'uncivilized' by Russell) is being corrected. A leading French anthropologist, Claude Lévi-Strauss, writes of their 'disinterested, attentive, fond and affectionate lore' concerning the animals and plants around them. He quotes comments by Canadian indigenes on the ignorance of visitors who have been only briefly in their country. The 'white man', they say, knows little about the animals; whereas they, the 'natives', know what the animals' habits are, what are the needs of the beaver, the bear, the salmon and others. Newcomers become confused when indigenous Canadians present their very real knowledge against a background of stories about the past: long ago, they say, their ancestors 'married' the animals and so learned their ways.

All human groups combine objective, practical knowledge with ancient myths. A famous example is the Egyptian 'Edwin Smith papyrus', from the seventeenth century BC, which appears to be a copy of an even earlier text. It deals accurately with anatomical and surgical matters but it also includes a portion on charms and magical practices. Familiar examples of ancient myths which are still influential can be found in *Genesis*, the first book of the Bible: there we have an account of a flood which probably happened, and the miraculous rescue of the animals of a large ecosystem.

Modern science, founded on the classifying and testing of the world begun by our ancestors, has become clearly separated from myth and magic only during the past four centuries. Today, foxglove leaves are known to contain cardiac glycosides which influence the action of the heart: among them are digitoxin and digoxin. These take us from folk medicine to complicated, life-saving chemistry and physiology. In some countries, people live more than twice as long as they did a century ago: many (among them, the author) are alive and active owing to use of drugs originally derived from plants or fungi. Other substances provide millions with relief from chronic pain. But the original, unsophisticated finding is still valid. If the discoverer began with a hypothesis, then the finding was scientific in a quite formal sense.

PSYCHIC MAGIC

We are therefore sometimes said to live in a scientific world. But in fact Russell's comments on 'unfounded beliefs' should not have been confined to tribal people. Plenty of nonscience continues to make obstacles to understanding. In 1997, a satirical newspaper columnist announced that, on 31 December 1999, as a preliminary to the Olympic Games, the 'Psychic Spiritual Olympic Grand Finals' were to be held in Canberra. The thrilling program included walking on water as well as the usual telekinesis and animal telepathy. The winners would get gold medals. Presumably the gold would be produced by alchemy. In the world of the occult, one can acquire gold, or develop some useful ability, without exertion. One gets something for nothing: no difficult mining or refining of ores; no blood, toil, tears or sweat at the laboratory bench. Certainly, no science, only magic.

The reader may suspect that the enthusiasts for 'Paranormal Olympics' are just a tiny group of eccentrics—or, more likely, that they do not exist. Yet a study of Canadian university students found a majority to believe in three major items from the paranormal repertoire: astrology, reincarnation and extrasensory perception (or ESP—which includes telepathy).

Similarly, in Australia, in 1987, Vernon Tupper and Robert Williams questioned 161 undergraduate students of psychology. About forty per cent accepted the astrological notion that personal character is influenced by the planets. Evidently a higher education does not interfere with liking for the occult. (For more, see C.E.M. Hansel on ESP.)

In America and elsewhere the heavens are watched for flying saucers or UFOs. In the USA, one out of every four adults believes that aliens have landed on earth. Many Americans are convinced that they have been kidnapped, while asleep, by extraterrestrials, inspected and returned—thrown back, you might say. The extraterrestrials are commonly sexually active, which suggests wishful thinking as one source of these stories; but a support group has been set up for people who say that they have been raped by alien visitors. Unfortunately, despite many attempts, no convincing photographs of the aliens have yet been published.

Similar beliefs in the paranormal have been revealed by systematic enquiry in other countries, including Britain, Germany, Ghana and India, and among older people as well as students. When scepticism is expressed, it may be attributed to jealousy on the part of bigoted scientists helpless to explain phenomena outside their scope.

In 1988, Australia made a special contribution to the paranormal, for there 'Carlos' made his sensational tour. 'Carlos', a two thousand-year-old spirit, intermittently entered the body of a young American who then fell into a deep trance. Through him, cheering messages were uttered. The spirit had plenty of sympathetic coverage in the media; but, when the host of one broadcast program dared to ask sceptical questions, 'Carlos' put him under a curse. This caused consternation: dire warnings were uttered by believers. A stage show, however, before a large audience, was a great success.

Then a TV program revealed the whole thing as a hoax concocted by a famous American stage magician and sceptic, James Randi. His dupes, in the media, were furious.

The experiment with 'Carlos' may be regarded as a contribution to research in social psychology. It tested a hypothesis: that many people can be fooled, including some in responsible positions. The findings matched the hypothesis. The experiment also brought out something quite typical of psychics. Even if they refrain from uttering maledictions on unbelievers, they permit investigation only in conditions chosen by them. Their achievements are announced with authority and must be taken on faith. In this, they are the opposite of science.

Even in the academic world, scientific rejection of authority has been resented. When, in the nineteenth century, laboratories for teaching science were proposed in Cambridge University, a professor objected: undergraduates, he said, need not see experiments, for the results could be guaranteed by their teachers, all of whom were persons of the highest character. Bertrand Russell, who tells this story, adds that many of them were, moreover, clergymen of the Church of England.

But, eventually, as we see throughout this book, dogma and arbitrary authority fail to meet the tests applied to them.

PROPHETS, PROFITS AND INKLINGS

In 1998 an English professor of the history and sociology of religion, Peter Clark, was reported as saying that the late twentieth century is the most superstitious age on record. He had good reason for saying so. In May 1998, after this chapter had been drafted, a 'Mind–Body–Spirit Festival' was held in London. Those who attended were told that they could, among other things, have an out-of-body experience, release their psychic energy and have their auras photographed.

It was a profitable venture with plenty of customers. About two-thirds of the British population believe in the paranormal and a similar proportion accept astrology. Other important beliefs, however, are less popular: a European survey has found only 22 per cent of adults accepting reincarnation.

An academic writer has predictably weighed in and stated that a 'religious impulse' is genetically programmed in humanity as a protection against the knowledge of death. One wonders how he knows. By instinct, perhaps.

In 1999 another academic writer, in a journal addressed to his colleagues and pupils, asked 'whether we humans have an inkling of the future direction of things'. He describes how a friend (said to be a crusty forester in his sixties) was driving in traffic, when his wife next to him cried out that the baby in the car in front was about to fall out. The car in fact showed nothing unusual; but a few seconds later it made a turn, a door opened and a baby fell on to the road.

The cautious reader may want to know something more about the driver (was he a fantasiser?) and his wife (did she often cry out like that?), and also about the person reporting the story; still more important, were there other witnesses? But the leading obstacle to scientific comment is the singularity of the event. It would be difficult for an experimenter to arrange for a series of such incidents.

My comments on this story make a statement on what, in science, is accepted as valid evidence. It would be irresponsible for us to lower our guard in the face of anecdotes which cannot be repeated or even checked.

STELLAR MAGIC

Testing the occult has to be energetic and resourceful, for preferences concerning the paranormal change with time. As Carl Sagan has described, before the days of UFOs, incubi were popular—alarming beings who mount you, when you are in bed, and do frightful things. Nowadays (unless we include extraterrestrial rapists) we hardly hear about them. Nor do we often hear of the elves, formerly numerous—especially in wet, low-lying districts—who shot people with darts of disease. So we have fashions in fairy tales.

Some tall stories, however, have survived for millennia. The Greek hero, Ulysses, in Shakespeare's *Troilus and Cressida*, tells Agamemnon how, in orderly times, the planets

> Observe degree, priority, and place,
> Insisture, course, proportion, season, form.

But when the planets

> In evil mixture to disorder wander,
> What plagues and what portents, what mutiny,
> What raging of the sea, shaking of earth,
> Commotion in the winds, frights, changes, horrors,
> Divert and crack, rend and deracinate
> The unity and married calm of states
> Quite from their fixture.

The magic of heavenly bodies and animals, combined in an Egyptian painting of about 1300BC. Astrology is still influential even in our own day; but it rarely takes such a decorative form. (From L.T. Hogben & M. Neurath, *From Cave Painting to Comic Strip*.)

The form of astrology adopted in Europe emerged in Mesopotamia in the seventh century BC. It became a feature even in the affairs of that headquarters of the intellect, ancient Greece. In classical Rome, although prosecuted, it influenced the policies of emperors. (The word *influence*, derived from Latin, itself originally signified 'emanation from the stars'.)

Astrologers made precise records of the most readily observed heavenly bodies, especially the sun, the moon and five planets—Saturn, Jupiter, Mars, Venus and Mercury. Unlike what goes on around us on earth, their movements, though complex, are predictable. Astrological studies were therefore, in some aspects, scientific: they encouraged the growth of mathematics and helped to found modern astronomy.

Each planetary name, however, is of a classical deity: among them, Mars stands for war, Venus for love, Mercury for trade. Correspondingly, the positions of the planets in the sky were held to influence, even to predict, wars and other political events; and the pattern of the heavens at the moment of birth (the horoscope) was believed to identify a person's abilities and prospects, including the outcome of legacies. Everyday speech still echoes these ideas. A volatile, optimistic, ready-witted person, we say, is mercurial. Joviality we associate with Jupiter, the largest planet and the father of the Roman gods: to be born under Jupiter was held to make a person joyful and happy.

In the European middle ages and later, astrology pervaded

society. Herbs were of great importance both for medicine and for cooking, hence some were planetary plants. A reader who still adheres to this doctrine should gather hazelnuts and olives only on Sundays.

Alchemy, which preceded modern chemistry, was more serious and in part an authentic study of metals. It also included many attempts to convert 'base metals', such as iron, into gold. Each metal was under the aegis of a heavenly body. An alchemist would have carried out experiments on the element tin (Sn, from the Latin *stannum*), only when Jupiter was in the ascendant. Eventually, however, when chemists decided to ignore the gods, they found their experiments to be unaffected.

Scepticism in other social circles came later. At the beginning of the sixteenth century, Antwerp in the Netherlands became a centre of trade of unprecedented power. Financial speculations, however, were precarious and financiers, then as now, mercurial. Hence astrological prediction flourished. Attempts were made to include astrology in the curricula of Dutch schools and universities.

Belief in the occult is especially influential in times of 'dire combustion and confused events'. That phrase, addressed to Shakespeare's Macbeth, can also be applied to our own period. Yet, according to the historian Keith Thomas, as early as the seventeenth century astrology had ceased, in all but the most unsophisticated circles, to be regarded as either a science or a crime: it had become 'simply a joke'. He tells us how it had been inadvertently put to the test among prominent persons. In 1669, the French king appointed an astrologer, the Abbé Pregnani, as his agent in England. Pregnani duly attended the horse races at Newmarket but, despite his efforts, the English king ended with no winners. This deplorable diplomatic incident was, in a sense, a scientific test of a hypothesis. The hypothesis was disconfirmed.

Not long before that royal fiasco, Francis Bacon (Lord Verulam, 1561–1626), statesman, lawyer and philosopher of science, wrote this.

> Atheism leaves a man to sense, to philosophy, to natural piety, to laws, to reputation; . . . but superstition dismounts all these, and erecteth an absolute monarchy in the minds of men.

Science deposes that absolute monarchy. Many tests of astrology, more elaborate than Pregnani's, have given negative results. Modern astrologers cannot predict a person's fate or qualities: they fail, for instance, when tested on twins—who are born 'under the same stars'.

A French investigator recently discovered the times and places of birth of ten men who had been convicted of especially horrible crimes. The diagnoses and predictions in their horoscopes did not match their actual lives: one Marcel Petiot had committed 63 murders for gain. So an offer was advertised of a complete, free, ultrapersonal horoscope; and each of 150 people, who responded with their date, time and place of birth, received a copy of Petiot's horoscope. Here it is, abbreviated:

> As he is a Virgo-Jovian, instinctive warmth or power is allied with the resources of the intellect, lucidity, wit. He may appear as someone who submits himself to social norms, fond of property and endowed with a comforting moral sense . . . that of a worthy, right thinking citizen. His affections toward others find their expression in total devotion to others, redeeming love, or altruistic sacrifices.

Of those who replied, 94 per cent recognised themselves in this portrait.

A vast audience, including some prominent persons, is in fact still willing to welcome the paranormal. Astrology remains a thriving occupation. The USA is said to have about 10 000 people who earn a living from horoscopes generated by computers. When Ronald Reagan was President of the United States, his wife, in the tradition of the Roman Emperors, regularly consulted an astrologer: this person was required to advise on which days were propitious for the great man to set out from home. Similarly, on 9 April 1998, the Indian Prime Minister A.B. Vajpayee, who had just taken office, was reported as waiting for an auspicious time, to be recommended by his astrologer, to move into his official residence. So astrology, though commonly derided, has not yet become *merely* a joke. (For more, see *Science and the Paranormal* by Abell & Singer.)

FICTIONAL MAGIC

Scientific theories and explanations are always open to being refuted. K.R. Popper (1902–1994), the philosopher who has especially emphasised the importance of refutation, has written on the growth of scientific knowledge:

> Science is one of the very few human activities in which errors are systematically criticized and fairly often, in time, corrected. This is why we can say that, in science, we often learn from our mistakes.

Astrologers, psychics and occultists, like some scientists, make predictions; but they do not learn from their mistakes. With Shakespeare's Glendower, in *Henry IV*, *Part 1*, they say, in effect,

> I can call spirits from the vasty deep.

But Hotspur responds,

> Why so can I, or so can any man,
> But will they come, when you do call for them?

Hotspur's sceptical response is the sort of obvious, naive question asked by scientists. And we may ask: if they do come, what will they tell us? Their usual offerings, like those of 'Carlos', are soothing messages and moralities. With many others, I should very much like to know how Mozart played the slow movements of his piano concerti, but that is not the sort of thing they divulge.

Among psychic phenomena, clairvoyance and telepathy are especially popular. A reader who enjoys thrillers may have been impressed by how often these abilities appear in well made, modern crime stories in which the author goes to massive pains to get technicalities right. In books by a successful American crime writer, Patricia Cornwell, the heroine is a forensic pathologist and a lawyer who also uses computers in the modern style. The pathology, the law and the computing are carefully authentic. But, at one point, in *All that Remains*, the heroine dashes off to consult a clairvoyant. This person, an elderly, rather witchlike woman, says that she possesses enhanced intuition. Given a photograph, she rubs her fingers over it and is able to say whether the person in the picture is alive or dead; she also detects emotions associated with the person.

In the everyday world we may ask what happens when clairvoyant abilities are rigorously tested. Susan Blackmore, an English parapsychologist, describes how she has a five-digit number, a word and a small object, each concealed in her kitchen. The place and the kinds of item had been chosen by a young man who intended to identify them while travelling out of his body. But, after three years, he had still failed to identify even one.

Perhaps telepathy is more credible—direct communication from mind to mind: no speech; no gestures; no body language; not even e-mail. For this, we go to another writer of fully researched thrillers, Dick Francis. In two of his books, *Break In* and *Bolt*, he has a jockey hero with telepathic powers. One of the characters says that quite often the hero answers a question before it is asked. Asked how he does it, the jockey replies that he does not know.

So we return to Susan Blackmore. A mother and daughter from Scotland asserted that they could pick up images from each other's minds. When tested, they chose to use playing cards because they used them at home. They also chose the room in which they would be tested. Susan Blackmore, however, ensured that the 'receiver' could not see the cards by any normal means. They failed. They were utterly disconcerted, for they had honestly believed that they could do it. Such negative results are typical of what happens when the paranormal is carefully studied.

Modern novelists enticed by the paranormal should consult Conan Doyle's Sherlock Holmes. In *The Resident Patient*, Dr Watson, in London, is bored, on a rainy October day.

> Finding that Holmes was too absorbed for conversation, I had tossed aside the paper, and, leaning back in my chair, I fell into a brown study. Suddenly my companion's voice broke in on my thoughts.
>
> 'You are right, Watson. It does seem a very preposterous way of settling a dispute.'
>
> 'Most preposterous!', I exclaimed, and then, suddenly realising how he had echoed my inmost thoughts, I sat up in my chair and stared at him.

'What is this, Holmes? This is beyond anything I could have imagined.'

To which Holmes responds:

'Some little time ago, I read you a passage from Poe in which a close reasoner follows the unspoken thoughts of his companion. You were incredulous . . . But the features are given to man as the means by which he shall express his emotions, and yours are faithful servants.'

And then Holmes explains at length how he watched Watson's face and movements, and followed his train of thought.

Sherlock Holmes was right. We can all interpret the features and the body language of others, especially those of people we know well. We observe the complicated muscles in their faces—the muscles of expression; the whites of their eyes enable us to see their direction of gaze and their red lips make lip reading possible. This is not extrasensory: it is straight perception. Some people are unnervingly good at it.

An American psychologist, Paul Ekman, has made a systematic study of the human eyebrows: how they express our feelings, often without our knowing it, and how we can interpret the eyebrows of others. The forehead, he finds, is the chief site of informative muscle movements. The most used facial expressions are raising and lowering the eyebrows. These are used during conversation to accent or to emphasize speech and as question or exclamation marks. Certain muscles pull the brows down and together. Charles Darwin called them the 'muscles of difficulty', for this action occurs with difficulty of any kind, from lifting something heavy to solving a problem in mental arithmetic.

In grief and other kinds of distress, the inner corners of the eyebrow are pulled up. When people are asked to make this movement deliberately, fewer than fifteen per cent succeed. It therefore does not appear during a false display of these emotions; but it does occur when a person feels sad even if an attempt is made to conceal the feeling. Similarly, in fear or anxiety the eyebrows are raised and pulled together; this combination of actions can rarely be achieved at will.

ANIMAL MAGIC

Ekman's work illustrates a general principle: anything we can repeatedly observe can be studied scientifically. Such studies, even when they are of familiar things, can yield unexpected findings.

Among them, our diverse relationships with animals, especially with our pets, can tell us much about both magic and science and, as a result, about ourselves. In ancient Egypt, cats were sacred. Families went into mourning when their cat died. In ancient Greece and Rome, conspicuous consumption by the wealthy and powerful included the possession of pampered lapdogs. In the courts of Chinese emperors, Pekinese were fed at the breasts of human wet nurses. In the European middle ages, bishops' palaces, nunneries and noble houses were homes to menageries of many species.

Yet, later, keeping pets came to be regarded as immoral: dogs were held to represent bestiality and even subversion. In sixteenth century England, according to Keith Thomas, a witch 'was likely to possess a familiar imp or devil, who would take the shape of an animal, usually a cat or dog'. Some witches were accused of copulating with the Devil in the form of a cat.

In many societies, combined magical and medical abilities have been attributed to animals. One story, preserved in ancient Greek writings, concerns a man with a sore toe who fell asleep outside a temple. A tame snake came out and licked the toe. He woke up cured. The Greeks also had temple dogs which licked and cured an affected part while the patient was awake.

Today, as far as I know, physicians do not prescribe the saliva of snakes or dogs; but a widespread notion does exist that, for some people, association with pets is beneficial for their health. We need not dismiss such statements as fantasies for, as S.A. Corson and A.N. Rowan and others have shown, they can be systematically tested. As usual, problems of method arise. The pets may be introduced with enthusiasm by nurses: perhaps the nurses, not the animals, are responsible for the improvement. Or the pets could be merely a source of extra stimulation: simply watching tropical fish swimming in an aquarium can be therapeutic. Tropical fish cannot be petted. These are examples of the need for controls: to make out what is happening, only one condition should be varied at a time.

Nonetheless, careful studies, which allow for such complications, have found favourable effects of making pets available in mental hospitals, homes for the aged and prisons. Greatest success is said to be among people who feel isolated or alienated from society. Ownership of a pet has also been said to favour both survival after a heart attack and recovery after surgery. Hence a plausible argument, based on scientific study, can be made for domestic animals as a source of wellbeing and a valuable adjunct to therapy; and pets are increasingly encouraged, or at least allowed, in many institutions.

Other research findings have, for some animal lovers, been less welcome. Some pet owners believe that they have special ways of communicating with their cats, dogs and horses. Animals do indeed respond to people in complex ways. The Elberfeld horses, trained in Germany and Austria early in the twentieth century, were among those said to possess extrasensory powers or to be superintelligent. The best studied was Clever Hans, owned by a German teacher of mathematics, Wilhelm von Osten, a gentle and clever trainer who said that he used methods derived from the classroom. He was convinced that his horses were much more intelligent than most people believed.

Clever Hans himself seems to have been inattentive, even bad tempered. But, when interrogated in German, he could answer simple questions, some of them arithmetical, by stamping a hoof or by pointing with his head. Hence he and his owner became famous.

Some people, however, raised their eyebrows. They set up a powerful Commission which, after critical investigation, was able to rule out fraud. A crucial finding was that Clever Hans could perform in the absence of his trainer. How did the horse do it? A professor of psychology gave the problem to a graduate student, who did a thorough job of research. He asked how von Osten trained his horses, and then did experiments which produced long tables of figures. These revealed that the horse could cope with questions only in the presence of somebody who knew the answers. Clever Hans had learned to respond to cues given unconsciously by observers. Usually, the cues were movements of the head. It was helpful if the watchers wore hats with broad brims.

The 'Clever Hans error' has therefore become a standard cautionary tale for experimenters on animal behaviour. It shows the need for taking elaborate care to avoid giving unintended information to experimental subjects.

An additional reason exists for including this story. A much discussed book by two journalists, William Broad and Nicholas Wade, has the title, *Betrayers of the Truth*. It was published by a major university press and purports to be an exposure of scientific fraud. But, amazingly, the authors include the case of Clever Hans, when the central feature of the achievements of the Elberfeld horses is that they are *not* fraudulent. So we have to contend, not only with irrational beliefs and exacting problems of method, but also with pointless misrepresentations by people who should know better.

MEDICAL MAGIC

Like astrology and astronomy, magic and rational medicine began to separate only after a struggle. Not long ago, medical practice included necklaces to ward off evil spirits; or a physician, in order to improve a patient's chances of recovery, might prescribe a change of name. In the nineteenth century, in rural France, it was believed that severe vomiting resulted from the stomach becoming unhooked and falling down. So the physician himself would go into contortions to unhook *his* stomach and then hook it up again. The patient recovered. Fee: five francs.

Magical medicine had a tenacious hold also in England. A historian of science, Charles Singer (1876–1960), tells of a Justice Holt who, before he qualified in law, was a wild youth. On one occasion, penniless near Oxford, he paid for a week's lodging by posing as an apothecary and treating the landlady's feverish daughter: he wrote Greek words on parchment, rolled it up and ordered that it be tied to the girl's wrist until she recovered. Many years later, an old woman, who regularly treated ague (or fever) with a magical parchment, was charged in his court with sorcery. The charm proved to be the Judge's own fragment, well preserved. The Judge confessed and freed the prisoner. We are not told what effect, if any, this had on the local superstitions; but the woman was one of the last to be tried for witchcraft in England.

Modern medical science has given us methods of preventing or treating disease which appear magical: for instance, vitamins, hormones and antibiotics. Sometimes, no doubt, a doctor's written prescription seems like a charm; but the medicines prescribed are (usually) the outcome not of sorcery but of the patient grind of systematic research.

Like the investigations of calculating horses and therapeutic pets, a central feature of such research is the control experiment. A sleepless patient, let us say, is injected with a drug believed to be hypnotic. The patient duly falls asleep. Evidently, the drug works. But does it? The patient knows that a drug has been given and that it is supposed to induce sleep. Also, the nurse confidently says that it will. Can such assurances make a patient sleep? Certainly, the soothing or encouraging attitudes of physicians and nurses are therapeutic in themselves. Such effects of suggestion are part of a therapist's equipment.

To discover whether the drug is effective unaided, one must control for the experience of being injected and for being told that the drug will work. This requires the 'double blind'. Some subjects are given the drug. Others, the controls, receive an apparently identical medicine (a placebo) which contains no drug. The subjects and the nurses do not know which is which. The question then is: do the subjects, who receive the drug, respond differently from the controls given the placebo?

The double blind is not easy to do. In some experiments, the drug has obvious side effects, such as causing a dry mouth. If the placebos have no such effect, they can be identified. The controls therefore have to be given pills which dry up their mouths also. As well, the two groups, experimental and control, must be matched: the members must be similar in age, sex, social class and other features. This is difficult to arrange.

Rigorous enquiry not only identifies effective treatments but also helps to protect us from error, fraud and charlatanry. In the 1990s, a successful movie, *Lorenzo's Oil*, portrayed a small child with a rare, crippling disease of the nervous system, ALD (adrenoleukodystrophy). The parents search the medical journals and find a possible remedy. The plot could have presented the anguish of the parents and the harsh facts of coping with a disease with no known authentic treatment. But that would not have made good box office. An eminent American medical scientist, Michael Bishop, has written:

> The film portrays the treatment of Lorenzo as a success, with heroic parents triumphant over obstruction by medical scientists. The film says nothing of the many children with ALD who received the oils in controlled studies, without showing any convincing improvement. The film is deeply troubling in its portrayal of medical scientists as insensitive, close-minded and self-serving; and in its impatience with controlled studies as wasteful of time.

Late in the twentieth century, the power of suggestion and the dangers of quackery have been newly exposed by bizarre instances of psychotherapeutic malpractice. These cases also, eventually, revealed the value of scientific scrutiny. A young American woman consults a therapist who induces her to recover 'repressed memories' of terrifying experiences: under his guidance, she reports being repeatedly raped in childhood by her father, with the connivance of her mother. She also describes two resulting pregnancies, both aborted. Eventually, at the age of twenty-two, she is medically examined. She proves never to have been pregnant and, indeed, to be a virgin. Another woman, an American hospital nurse, consults a psychiatrist about her daughter. The therapist hypnotizes the nurse and, for a time, convinces her falsely that she is a case of multiple personality (dissociation); and that in childhood she had experienced hideous abuse.

Other such cases have been fully reported (and have resulted in the payment of massive damages by the perpetrators). In each, grossly improbable stories of early experiences were accepted but later denied; and the denials were conclusively shown to correspond to the facts.

One outcome has been research which exposes the contrast between science and untested speculation. Some people accepted the stories of early abuse because they seemed to match the theories of psychoanalysis. The originator of the theories, the Austrian physician, Sigmund Freud (1856–1939), held out his system as scientific; so did many of his followers; but critics have dismissed its main assertions as untestable, hence quite outside the natural sciences.

In psychoanalysis, the idea of repressed memories has been central. In his last work, *An Outline of Psychoanalysis*, Freud states that the sexual abuse of children by adults is common; that such experiences are not remembered but are actively dismissed from memory; and that the result is neurosis. He writes:

> The helpless ego fends off these problems by attempts at flight (by *repressions*), which turn out later to be ineffective and which involve permanent hindrances to further development.

Successful treatment of the neurosis, he says, requires that the repressed memories be restored to consciousness.

In a work, *Greatness and Limitations of Freud's Thought*, first published in 1980, a leading psychoanalytical writer, Erich Fromm (1900–1980), expresses doubts:

> . . . one of the crucial difficulties of Freud's assumption of the significance of early childhood [is that] much of what are supposed to be experiences of the earliest childhood are reconstructions. And these reconstructions are very unreliable.

Today, experimental evidence supports Fromm's scepticism. An American psychologist, E.F. Loftus, and her colleagues have devised ingenious methods by which disagreeable, invented memories can be implanted in adult volunteers. Some subjects were falsely told that, according to a relative, they had, at about the age of five, been frighteningly lost in a large, crowded building. In others, the 'memory' was of going to hospital with a high fever. They were also each told three true stories about their childhood. About one in three of the subjects, when questioned later, duly described not only the true events but also the implanted memories, as though they had actually happened. Not all the invented stories were disagreeable. One was of a birthday party with enjoyable scenes. These were similarly 'remembered'.

Hence the power of suggestion has been experimentally confirmed. Therapists have become more aware that many of their clients are highly suggestible and can be persuaded into attitudes or beliefs unintentionally supplied during therapy. The

psychoanalytic concept of repression is therefore being discarded.

The experiments also show why many critics hold Freud's
system to fall outside science. Psychoanalysis originated as a
method of treating patients, often seriously distressed, for whom
no established therapy existed. Some, no doubt, *had* been abused
in childhood. But Freudian conclusions, as Fromm implies, are
based not on direct, systematic observation, still less on experiment, but on inferences from the statements of patients and on
the interpretations of analysts. The latter may be regarded as
conjectures or guesses. In a scientific study, the next step would
be to make more definite proposals which could be tested. The
experiments just described are examples of how speculation can
be corrected by science.

CHAPTER 2

BRAIN WAVES

> . . . and in his brain,
> Which is as dry as the remainder biscuit
> After a voyage, he hath strange places cramm'd
> With observation, the which he vents
> In mangled forms.
>
> SHAKESPEARE, *As You Like It*

THE OCCULT MYTHS AND speculations which occupy much of the previous chapter are rarely thought of as scientific. The present chapter is a bridge between the still popular ancient magic and the modern myths and fables which have, or seem to have, a genuine foundation in science. Pseudoscientific stories, like astrology, are part of our social history. They can obstruct the understanding of science and of human action; but, seen closely, they can also reveal much about what scientists do.

BUMPS ON THE CRANIUM

Phrenology, a leading case of error, arose from study of one of the great mysteries: the localization of function in the brain. A German physician, F.J. Gall (1758–1828), saw the problem with (at first) impressive clarity. Gall practised medicine in Vienna, where he also made major contributions to knowledge of the nervous system. The cellular structure of tissues was not yet fully established. Nerves were still thought of as hollow tubes. Gall began with microscopic observations on nerve fibres which could be found in animals generally, even worms. He wrote accurate reports on ganglia, where many nerve fibres meet, and on the massive bundles of fibres in the spinal cord. From the spinal cord, millions of fibres lead to the brain. In order to

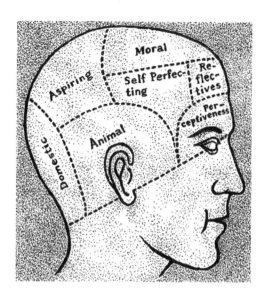

A phrenologist's entirely imaginary grouping of 'organs' in the brain. This diagram was published in 1869, long after the theories of phrenology had been refuted. (From C. Colbert; after O.S. Fowler, *The Practical Phrenologist*)

support his ideas about the relationship of brain structure with function, he also used comparisons of many species. And, for Gall, the brain was not only the site of intelligence: it was also the seat of the soul. At this point he got into trouble: the Viennese authorities condemned his views as excessively 'materialistic' and censored them. He therefore left Vienna.

So far, Gall appears as a notable figure in the history of neurology and perhaps as a minor martyr to obscurantism. But he now changes. He and an assistant, J.C. Spurzheim (1776–1832), travelled widely in Europe, where they attracted much attention with a new theory. The cerebral hemispheres, they said, contained many separate centres, each with its own function. The form of the skull matched the development of the centres. A person's 'innate' character and skills could be inferred from bumps on the cranium. A large bump in a certain place indicated conscience, another showed conjugal love, yet another, self esteem. The bigger the bump, the more pronounced was the trait. A bump held to indicate acquisitiveness was said to be especially large in pickpockets.

It was not difficult to refute phrenology (or 'craniology').

Charles Colbert, in a historical study, gives the example of an anatomist, Thomas Sewall, who in 1837 found none of the alleged divisions when he dissected actual brains; nor, he showed, could examination of the skull give accurate information on the shape of the brain.

As the historian, R.M. Young, writes, when confirmatory evidence for phrenological theories was needed, Gall 'had almost no standard of evidence'. Like some more recent romancers, he wrote extensively about the 'instinct to kill' of animals; but on this topic his discussion 'consists mostly of anecdotes about a particularly carnivorous lapdog'. In 'man' he found 'an innate propensity which leads him to destroy his own species'—another adumbration of twentieth century pop biology.

Gall, on no good evidence, was also convinced that the cerebellum is the centre of the 'reproductive instincts'. Here he came up against a pioneer of modern, scientific physiology and medicine, J.P.M. Flourens (1794–1867), a French physician and comparative anatomist, who emphasized the need for direct observation and well designed experiments. One of his lasting findings, based on his study of many species, was that the function of the cerebellum is coordinating bodily movements: an animal with a damaged cerebellum may be unable to walk without falling over.

Although Gall was a pioneer in neuroanatomy and in the study of human personality, his standing as a scientist was diminished by his wild theories. Phrenology was, however, helpful to charlatans: for many decades a phrenologist, like an astrologer with his horoscopes, could earn a good living with impressive maps of crania and assertions about his clients' characters. Gall's follower, Spurzheim, had great sucess when he toured the United States. He has been described as a propagandist rather than a scientist. If he lived today, he would have a fine time in the press and on television.

In another parallel to the present, phrenology during the nineteenth century had much influence among American artists and writers. Charles Colbert describes how it was little diminished by earlier refutations. Similarly, it was taken seriously by learned persons in England. One was A.R. Wallace (1823–1913), the joint originator with Darwin of the concept of natural selection. Another—for a time—was Herbert Spencer (1820–1903), an engineer whose writings for many years dominated sociology. For some, phrenology became a secular religion. It also passed into everyday language: people talked casually about their 'bump of direction', when they meant their ability to find the way about.

As we see later, such clinging to obvious error continues; but the errors are new ones.

THE ACTUAL BRAIN

During the nineteenth century, neurologists followed Gall's early example and produced extensive findings on the minutiae of neural structure. The staggering number of the (approximately) 100 000 000 000 nerve cells in the human brain was calculated and their diverse types were described. Many of their connexions were ingeniously mapped. From this work, certain conclusions on local function were obvious. Large tracts of fibres connect each of the main sense organs to distinct regions of the cerebral hemispheres. Hence we have the familiar charts of the primary projection areas: the auditory cortex receives information from the ear; the visual cortex—via the thalamus—from the eye and so on. In addition, the motor cortex sends fibres down the spinal cord and these stimulate nerve cells in the cord which connect to muscles. All such tracts are accompanied by others which carry impulses in the opposite direction.

The charts also show large 'centres' for understanding and producing speech. A French physician, J.B. Bouillaud (1796–1881), who had been taught by Gall, studied the effects of injury to the brain. As a result he identified a 'speech centre' in the middle of the left cerebral hemisphere.

Not much notice was taken of this finding, perhaps because Bouillaud supported phrenology. But, in 1861, P.P. Broca (1824–1880), a French surgeon and anthropologist, described a patient who had lost the ability to speak (aphasia), owing to damage to the left frontal lobe of the cerebrum (later called Broca's area). And, in 1874, Carl Wernicke (1848–1905), a German physician and anatomist, described a patient with diminished understanding of speech after injury to a region of the left temporal lobe (now Wernicke's area). These findings were especially striking, because speech is a central and unique

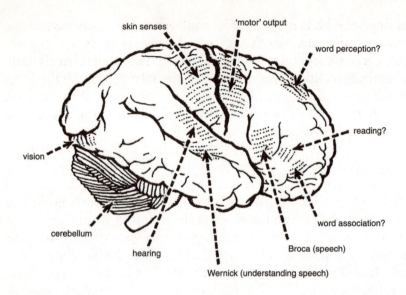

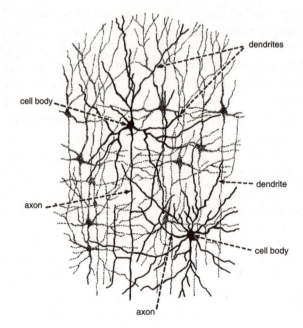

Top: a conventional (but misleading) drawing of a human brain seen from the right side. The labelled regions include the familiar ones and some, with question marks, which are derived from recent experiments described in the text. Does this diagram represent a questionable return to the long rejected ideas of phrenology? *Below*: nerve cells. Each cell interacts with hundreds of others; the human brain has thousands of millions of cells of these and other kinds. (Lower figure after J.Z. Young, *The Life of Mammals*)

trait of the human species. The two areas are, correspondingly, hardly represented in the brains of monkeys or apes.

Outside the cerebral cortex, experiments on animals have revealed localities with, evidently, special functions. Damage a small part of the hypothalamus, above the roof of the mouth, and the animal loses its appetite; another part, and the animal overeats and becomes obese.

Further up is the limbic system. 'Limbus' means a border. The limbic system forms a border round the brain stem where it joins the cerebrum. It has a complex structure which presented early anatomists with difficulties. One part is pear shaped and so is called the pyriform cortex; another is the amygdala or almond; a third, the hippocampus, was held to resemble that weird fish, the seahorse (*Hippocampus punctulatus*).

The front part of the limbic system receives an input from the olfactory organs; and, for some time, it was called the smell brain. Much of it, however, has intricate, two-way connexions with the hypothalamus below and the main part of the cerebrum above. Gently pinching the tail of a normal cat may evoke a faint sound; but, if the hippocampus is electrically stimulated as well, the result may be a violent seizure. Injuring this region can interfere with the ability of an animal to avoid a painful stimulus. Damage can also evoke wild behaviour *or* unusual tameness. Hence the limbic system is sometimes called the seat of the emotions; but such statements are not very informative.

Inevitably, in textbooks, the findings on structure are shown in simple diagrams. As well, novel discoveries are commonly described in terms already used for familiar things, such as 'pathways'. The brain has even been represented as a kind of telephone system. The expression 'hard wired' still appears in serious writings. Such use of analogy and metaphor, which we meet repeatedly in this book, is unavoidable. It is also dangerous. Applied to the nervous system, it can be grossly misleading about how the system works. Worse, it can encourage the notion that human beings are no more than robots or mechanisms to be tinkered with.

The inadequacy of telephone analogies was shown when the nerve impulse was analyzed. Signals pass along mammalian nerve fibres at speeds from one to 120 metres a second. Each impulse consists of a brief chemical change accompanied by an alteration of electrical potential in the fibre. The largest fibres, though more than a metre long, are only 20 μ (thousandths of a millimetre) in diameter; so for a long time it was difficult to detect what was happening in a single fibre.

In the 1930s, however, cephalopod molluscs gave some help. Small squids (*Sepia officinalis*), until then better known to gourmets, were found to have a few neurons whose axons are improbably large. A zoologist, J.Z. Young (1907–1997), describes how, when he first saw them, he thought they were veins. At their largest, these giant fibres are one millimetre in diameter. The impulses in them (action potentials) could be readily recorded with the equipment available. Here is an example of the fact that scientific advances often fail to follow a 'logical' progress.

Although research, at the level of cells and their chemistry, soon showed what happens in nerve fibres, it did not answer any of the important questions about brain function. It remained tempting to think of the brain as an assembly of wires and relays. But nerve fibres are not wires; nor are nerve fibres simply connected to other cells in a straightforward sequence. Each nerve cell in the cortex is in contact with many hundreds of others and each is related to *all* the others through at most four 'relays'. Every region of the cortex therefore has two-way connexions with every other. Similarly, cortical regions have reciprocal connexions with centres outside the cortex. Neither electrical engineering nor computer science can tell us how such a system works.

The great American psychologist K.S. Lashley (1890–1958), who rebelled against a crude reduction of the brain to a telephone system, summed up his life's work in the phrase, 'In search of the engram'. His primary question was: how does a brain enable us to learn and to remember? (He also once said that the phenomena of learning look impossible.) His experiments, mainly on domestic rats, seemed to show learning to be a function of the whole cerebral cortex: the ability to acquire and to remember new habits depended on the amount of cortical tissue present ('mass action'). He therefore proposed a holistic account of the cerebrum.

Yet, recently, much evidence has pointed the other way. Since the days of Broca and Wernicke, the results of wounds from war, other violence, accidents and experiments on animals have led to a renewed belief in minutely localized functions. When we see objects, or hear sentences and learn to understand them, we observe shapes, colours, sounds and other attributes as a combined experience; but, judged by some effects of local injury, in the brain each aspect seems to be 'processed' separately: damage to one small region may go with inability to distinguish colours (achromatopsia); to another, failure to understand certain words (semantic aphasia); to yet another, loss of the ability to name common objects (nominal aphasia). Some people lose the ability to write (agraphia), others, the ability to read (alexia). Many such special regions have been identified.

A particularly distressing but rare result of brain damage is prosopagnosia, or loss of our remarkable, and socially crucial, ability to recognise faces. This defect illustrates the complexity of brain function. The patient, shown a picture, cannot identify the person pictured but can say whether the person is (for instance) a politician or an actor. Correct identification depends on previous knowledge of the person.

Still stranger is 'blind sight'. Injury to the visual cortex at the back of the brain can result in loss of vision on one side: half the field of vision is blank. In carefully designed experiments, subjects with this defect can be 'shown' an object on the blind side and asked to reach out or to point to it. The subjects declare that they are only guessing; yet their guesses are, to their astonishment, usually accurate.

EXPERIMENTAL THINKING

The bizarre defects described by physicians and surgeons cannot tell us how the brain works. They are usually the result of indiscriminate destruction of millions of nerve cells and fibres. More important, if distinct regions do exist, each with a special role, we do not know how they combine to produce conscious experience and deliberate action. This has been called the central problem of brain physiology.

Today, however, neurologists need not depend on bashing the systems they are trying to understand. Among their new methods, PET (positron emission tomography) allows them to pick up local activity, even in the human brain, by detecting changes in blood flow. A volunteer is injected with a solution labelled with a radioactive isotope of oxygen. (The dose is small and the radioactivity decays in a few minutes; hence the procedure is held to be harmless.) Another method, magnetic resonance imaging (MRI), which gives more detailed findings, measures local changes of oxygen consumption. Both methods require very expensive equipment and the use of advanced computers.

A subject may be required to see a printed word; or to hear a word spoken; or to speak a word shown; or to make a verb from a noun. In each case a different pattern of blood flow is detected. When the subject has to hear a word, and then speak a different but related word, several regions become active; and the pattern changes when the subject has had practice at responding. Other experiments on cerebral activity have been done while the subject is doing sums. A large region of the prefrontal cortex then becomes active and also a smaller parietal region. These methods are, however, still crude and show cerebral action only on a large scale; and what the 'inactive' regions are doing meanwhile is an unanswered question.

Another enigma concerns the lack of differences between the brain cells of different regions. In other organs we find distinct regions with different functions. In the pancreas, most of the gland cells secrete enzymes which pass into the intestine and take part in digestion; but among them are groups of distinct cells (the islets of Langerhans): these secrete two hormones which pass into the blood and regulate sugar metabolism. In the cerebral cortex the microscope shows us many kinds of nerve cell but no islets corresponding to different functions: most regions have similar cells. These cells produce not secretions but—it seems—everything we call thought and feeling; we have no description at all of how they do it.

It is uncertain whether we should think of the cortex as divided into tidy, distinct regions, for two further reasons. First, some cerebral map makers have evidence of alteration in the invisible boundaries between regions owing to stimulation, that is, experience. Second, they have found variation between individuals in the localisation of functions, especially those concerned with language.

Localised or not, at the conscious level we have at least three kinds of memory: long term, short term and very brief or 'working' memory. The last is the kind we use to perform complex but momentary actions, such as uttering or typing a grammatical sentence. One has to 'remember' what has just been said or typed, in order to complete the sentence correctly.

Working memory has been studied by a method not applicable to human beings: electrodes are implanted in the brains of monkeys and changes are recorded while the animal is performing a set task—usually, recalling where it has seen an object a brief time before. This kind of ability is in constant use in the waking brain. It involves several cortical regions but especially the confusingly named prefrontal cortex just behind the forehead. A macaque with a damaged prefrontal cortex can perform delayed choice only with difficulty, if at all.

Compared with that of closely related species, the human prefrontal cortex is enormous. If findings on other species show it to be crucially concerned with memory, what happens, the reader may ask, if it is injured in a human being? Unfortunately, we know the answer. For some years in the mid twentieth century a small group of neurosurgeons carried out an operation, prefrontal leucotomy, on thousands of patients with a variety of disorders: they made cuts through this distinctively human structure. The procedure had no foundation in clinical research or in neurological science. It was an example of the willingness of a few people to regard people as mechanisms. Its effects were calamitous: they included deficiency in the ability to plan ahead and a generalised inertia.

In the book quoted above, J.Z. Young begins with the question, 'How does the brain work?' Of course, he does not answer it, or even pretend to do so. Since he wrote, modern experimental methods, such as PET and MRI, have yielded remarkable coloured pictures of brain activity; but no agreement has been reached on what they signify. Many workers hope that they will lead to new methods of treating disorders of the nervous system but in this little progress has been made. Some experimenters remain impressed by the evidence for strictly localised function. Others point to contradictory findings, especially the increasing amount of activity, dispersed in the cortex, when difficult problems have to be solved.

Despite the difficulties, a 'modular' account of neural

processes and thinking has become popular. The brain is seen as consisting of units (modules), each with a separate function. Stanislas Dehaene, in his admirably critical *The Number Sense*, writes:

> The very notion that 'thought' . . . can be localized in a small number of cerebral areas recalls an old discipline that was once relegated to the museum but is making an insidious comeback: Gall and Spurzheim's phrenology.

His comment may seem a little harsh, for the modern 'craniologists' do base some of their conclusions on careful experiments. But, more important, Dehaene points to a possible need for a partial return to the discarded 'holistic' hypothesis of K.S. Lashley.

So, after a century and a half, brain science is still immature. No dramatic breakthrough is in sight. Research on the brain is therefore in several respects typical of experimental biology: it consists of the patient, meticulous accumulation of detail, critical analysis of the results and much argument about what the results mean and what to do next. The arguments are, for the most part, decorous and rational. They are rarely directed to provoking headlines in the press. Much research on the brain is at the level of cell physiology or biochemistry: it then need not impinge directly on a reader's images of neighbours as social beings. In this too it is typical of most science. But, as we see in the next chapters, the accounts of the human sciences presented to the reading public are often different.

PART II

STRUGGLE

The adequate record of even the confusions of
our forebears may help, not only to clarify these
confusions, but to engender a salutary doubt
whether we are wholly immune from different
but equally great confusions.

A.O. LOVEJOY

The next four chapters are on consequences of the revolutions
in understanding founded by Charles Darwin (1809–1882) and
J.G. Mendel (1822–1884). Evolutionary biology and the new
genetics have, in little more than a century, revolutionised our
understanding of living things. They have therefore been used
in attempts to give scientific accounts of human society.

They have also been misused. The phrenological ideas of a
modern Joseph Gall, we may be sure, would be received enthusi-
astically by the modern media, especially because Gall proposed
an unfounded theory of hereditary intellectual qualities.

Yet even erroneous stories of human biology can be valuable.
They oblige us to look hard at our assumptions about our own
species and they can help us to decide on both the scope and
the limitations of scientific knowledge.

CHAPTER 3

APE OR ANGEL?

Is man an ape or an angel? Now I am on the
side of the angels.

BENJAMIN DISRAELI

IN THE NINETEENTH CENTURY, biological science created
an upheaval in thought with which humanity is still struggling
to come to terms. If humanity is a product of Darwinian
evolution, what conclusions may we draw from that? What may
usefully be called human nature? What are the political implications of Darwinism?

ADAPTING TO DARWIN

Organic evolution consists of change, much of it adaptive: that
is, of adjustment to circumstances, especially changed circumstances, over many generations. The reason for the caution in
the phrase, '*much of it* adaptive', appears later.

The modern theory of evolution was founded by two
Englishmen, Charles Darwin and A.R. Wallace. Both were
learned in natural history and knew much about adaptedness at
first hand. Until their time, in Christendom, the existence of a
Designer had been generally taken for granted. Instead, they
explained adaptedness by a concept which does not require
conscious design. Darwin wrote: 'I have called this principle, by
which each slight variation, if useful, is preserved, by the term
of Natural Selection.' Darwinism therefore seemed to replace

35

the theists' 'first cause' with an amoral, impersonal process which can be investigated objectively and critically.

Instead of more generalities, here are examples of adaptedness.

The structure of organisms often reminds us of the design of machines. (In the past, their seeming perfection was held out by some writers as irrefutable evidence of a Creator.) Before the second world war, aircraft engineers were famously inspired by the form of a gull's wings. This was reflected in the design of the Spitfire, the fighter which, during that war, played a major part in the Battle of Britain.

Similarly, when, during the same war, echolocation (sonar) was developed for detecting submarines, it incorporated principles like those used when, in complete darkness, bats (Chiroptera) identify their prey, many of which are night-flying moths. A bat's ability not to confuse a moth with other objects (such as raindrops) depends on generating sounds of suitably high pitch, emitted at a suitably high speed; the sounds are beamed from distorted nostrils ('horns') which, until recently, were thought to be organs of touch. Obstacles, which have different echoes, are avoided. All echoes are received by ears of which the external structures (the pinnae) are usually very large, very mobile and of elaborate shape. An American zoologist, D.R. Griffin, has described this assembly of unique features, and much more, in a superb book, *Listening in the Dark*. He writes that a bat scans its environment with a beam of sound, much as a human being explores the dark with a flashlight. (He also describes how blind persons, by tapping a stick, use an ability, similar to that of bats, to identify nearby objects.)

Human beings, at least when young, can hear the squeaks of bats as they fly at night. Other animals respond to stimuli which we detect only by instruments. The precision and economy with which the honey bee (*Apis mellifera*) builds its combs could not be bettered by a human geometer. In a hollow tree, in the dark, the combs of a swarm are precisely parallel and have the same orientation as those from which the swarm originated. In an earlier period, they would have been said to achieve this 'by instinct'—an expression which suggests something magical but explains nothing. But, if a powerful magnet is switched on outside the hive and so deflects the earth's magnetic field, the orientation of the combs is altered correspondingly. After that astonishing finding, magnetic material was found in bees' abdomens.

Among other improbable abilities are the migrations of birds. Some perform barely credible flights twice a year. The legendary white stork (*Ciconia ciconia*) rears its nestlings in central and southern Europe but spends the northern winter in Africa, south of the equator. Arctic terns (*Sterna macrura*) nest in the Canadian Arctic but avoid the Canadian winter by migrating to the Antarctic. The Pacific golden plover (*Pluvialis dominica*) breeds in Siberia and Alaska but spends the northern winter in south-east Asia and Pacific islands.

The uncanny abilities of a small, common bird, the garden warbler (*Sylvia borin*), have been intensively studied. In autumn it flies, without guidance from its parents, from North Germany and other European breeding grounds to the Straits of Gibraltar. There it changes direction toward its central African wintering places. How does it achieve this feat of navigation, first southwest, then southeast? Experiments have shown it to make use both of the patterns of the stars and also of the earth's magnetic field. Such migratory species avoid not only cold but also dark days and so always have plenty of light to feed by. This is believed to be the survival value of their navigational abilities and their ability to fly enormous distances.

The more closely organisms are examined, the more we find adaptive complexities. The skeleton of insects is made largely of chitin, a substance rather like cellulose; it is usually combined with a protein and is often mixed with a hardener such as calcium carbonate. After a moult, the animal's epidermal cells secrete chitin as a fluid which flows out to cover the surface of the body and limbs. It then quickly hardens. It has been described as a typical plastic; but it antedates industrial plastics by more than five hundred million years.

Industrial chemists are still more impressed by spiders' webs. The silk of an orb weaver is a mechanical marvel, apparently flimsy but of astonishing strength for its weight: though elastic, its breaking strain is much superior to that of steel and it is tougher than rubber. The orb of an orb web species may contain thirty metres of silk. The silk is secreted as a watery mixture, yet the final product is insoluble (hence is not dissolved by rain). The mechanical properties of the silk depend on an alternation of orderly and disorderly arrangements of protein molecules. Some spiders secrete several kinds of silk, each with different properties. One may protect eggs, another may bind up the prey.

Spiders' webs have long been used as a dressing for wounds. Efforts are now being made to manufacture spider silk: when the genes of certain bacteria are suitably 'engineered', they can produce it in the form of granules. Many applications are expected, for instance in surgery and for body armour.

The squeaks of bats and the webs of spiders adapt their possessors to particular ways of living, but only in the context of other features. A bat's sonar would be useless if the bat were not also adapted for flying, with all the necessary features of structure and physiology. Each species of organism has its unique features. Thousands have been studied in detail. Much, though not all, of the unending and sometimes unexpected variety we see around us reflects adaptedness. As I look up, I see plants thriving in pots in the unpromising environment of my study; they remind me that, despite their need for light, some small plants can flourish at ground level in the gloom of a tropical forest. The reader can think of other examples. All are products of eons of change.

The concept of natural selection explains adaptedness by examining differences between individuals. Each complex living being differs from all others, even its near relations. (This is one reason why biology differs sharply from the physical sciences. A molecule of, say, table salt, sodium chloride, NaCl, is like any other.) Some individual differences are genetically determined and some genetical types have more descendants than have others. A kind of competition therefore seems to exist between different types.

Descriptions of this 'competition' have conventionally used other metaphors from human action, as in the expression, the *war* of nature. In Darwin's time, a slogan, 'the survival of the fittest', with its implication of incessant struggle and even enmity, came into everyday use.

The statements above, however, do not, at first sight, point to evolutionary change: on the contrary, they suggest stability. Each species is conspicuously adapted to its environment and way of living. Departure from the typical should therefore be disadvantageous. And, often, so it is. We can see stabilising selection in our own species. Although differences in human birth weight depend largely on the environment offered by the uterus, they are also genetically influenced. Death rates soon after birth are highest among the smallest and largest babies: hence selection evidently occurs in favour of near typical or average weight. Similar findings have been made on stature and the length of the menstrual cycle. For survival, it seems, it is usually helpful to be ordinary.

Darwin and Wallace, however, saw differences in survival rates also as a source of gradual change which could transform species. They achieved this inspired insight without direct evidence of progressive modifications in nature. But they did know that stockbreeders and farmers alter domestic animals and plants, over generations, by *selecting* those with useful features. The expression, natural selection, therefore, includes a metaphor provided by agriculture.

More of natural selection later. Before that, we examine the social impact of Darwinism—a collision which still has many echoes in our own day.

The 'Beast in Man': *Homo pugnax*

At the end of the first edition of *On the Origin of Species*, Darwin wrote:

> In the distant future I see open fields for far more important researches. Psychology will be based on a new foundation, that of the necessary acquirement of each mental power and capacity by gradation. Light will be thrown on the origin of man and his history.

The future foreseen by Darwin was not distant. From the 1860s, strenuous efforts have been made to interpret the human condition (comedy or tragedy as the reader prefers) by some derivative of Darwinism. These exertions have concentrated on our social behaviour.

Yet, in view of the importance of adaptedness, the question which should first be asked is: for what is *Homo sapiens* adapted? What, for instance, are the human traits analogous to the navigational abilities of birds or the architectural achievements of bees? One difference is obvious at once. Unlike those of animals, human skills are socially acquired and are often confined to specialists. A human being has to learn navigation gradually and only some need to do so. The learner requires intelligence and motivation and also a teacher, or at least a model to imitate. Similarly, builders and architects begin as apprentices. The adaptedness of our behaviour does not depend on special, fixed abilities (sometimes called instincts): it arises from adaptability. Much of our learning rests on cultural influences, especially teaching. Our modern skills are products of social, not evolutionary changes.

Though obvious, these features of humanity have often been disregarded. For a century and a half, 'Darwinian' accounts of our social lives have misled a large public by focusing on antisocial violence and, later, on sex and breeding.

> In the Neolithic Age savage warfare did I wage
>> For food and fame and woolly horses' pelt . . .
> But a rival of Solutré told the tribe my style was outré—
>> 'Neath a tomahawk, of diorite, he fell.
> And I left my views on Art, barbed and tanged, below the heart
>> Of a mammothistic etcher at Grenelle.
> Then I stripped them scalp from skull, and my hunting dogs
>>> fed full,
>> And their teeth I threaded neatly on a thong;
> And I wiped my mouth and said, 'It is well that they are dead,
>> For I know my work was right and theirs was wrong'.

In this way, in 1895, Rudyard Kipling (1865–1936), an immensely popular English versifier, reflected with cheerful brutality a common presumption of his time: that men, even practitioners of the fine arts, are belligerent oafs descended from cannibalistic cave dwellers.

This belief was combined with the longstanding practice of likening human beings to animals. For decades, as a result, 'war biologists' clashed with 'peace biologists'. The latter represented us not as a combative species but as one with an instinct for friendship (perhaps *Homo amans*). Hence the two sides used the same form of argument but drew opposite conclusions. All searched human social action for fixed evolutionary features (or instincts) typical of our species; their conclusions, however, reflected (and continue to reflect) not scientific findings but fantasies, wishes and fears concerning humanity.

Homo pugnax on a Greek vase from about 540BC. Achilles kills Penthesileia, Queen of the Amazons.

The principal resulting portrait, still influential, I have named *Homo pugnax*. A prominent English political scientist, Walter Bagehot (1826–1877), wrote in 1872 on applying natural selection to humanity. He was confident of the predominance, in human history, of the struggle for survival. Human progress, he held, depends on continuous strife: the strong prevail over the weak and the strongest are, in most respects, the best. This principle he applied both to individuals and to nations. His own metaphor for the struggle was that of a competitive examination. (He had, of course, himself done well in the examination room.)

Bagehot took for granted the superiority of the 'white race' in the struggle for existence: for him, 'primitive' or 'savage' peoples represented an earlier stage of evolution. He has been called (with some justification) a leading example of Victorian middle class smugness.

Sociology in the United States and Britain was at that time dominated by Herbert Spencer (1820–1903), the prolific writer who invented the slogan 'the survival of the fittest'. In 1870, in a widely read popular work, he too proposed a 'Darwinian' account of human society based on strife. Spencer regarded the 'cosmic process' of evolution with religious awe. He therefore regarded all measures of state welfare as acting against nature. But his writings, like those of his modern successors, lacked coherence. His opposition to governmental intrusions matched the outlook of extremist admirers of market forces. Yet he was personally benign; and he seemed also to look forward to a perfect state in which everyone was respected and in work. A historian, R.J. Richards, calls this vision 'a socialist utopia achieved through an evolutionary process'. According to this interpretation, Spencer held 'socialism' to be inevitable, so long as nothing was done to achieve it.

As an Australian historian, D.P. Crook, has shown, the extent to which early Darwinism directly influenced politics and war is debatable; but, since the 1870s, political ideas ostensibly based on Darwinism have had a remarkable scope and, near the end of the twentieth century, have become more prominent.

In Germany, Spencer was followed by Ernst Haeckel (1834–1919), a leading biologist whose interpretations of humanity brought him a large following. Haeckel, too, regarded natural selection as a cosmic force equivalent to a deity. He presented merciless conflict as a central feature of human action in which the white, Germanic races were certain to triumph. Negroes and Japanese were incapable of achieving civilisation. Jews were beyond the pale.

In this way, Spencer, Haeckel and others created 'social Darwinism'. Later, their ideas were echoed in writings which supported two of the leading criminal movements of twentieth century politics—Italian fascism and German National Socialism (Nazism). Benito Mussolini (1883–1945), founder of the fascist movement, rejected the idea of perpetual peace: war, he held, puts the stamp of nobility on all who have the courage to meet it. And Adolf Hitler (1889–1945), in his *Mein Kampf*, imagined a Nordic Race which, owing to its prowess in conquest, was 'the bearer of a higher ethic'. For him, it was a 'law of nature' that the strongest people should prevail.

The success of social Darwinism was achieved against resistance; for, as soon as it was founded, it was opposed. An early critic was T.H. Huxley (1825–1895), Darwin's most prominent supporter and a leading comparative anatomist. In 1893, Huxley's mature views were expressed in a famous lecture, 'Evolution and Ethics'. As he shows, in a long, careful argument, the fundamental error of Bagehot and others was what is now called biological naturalism—the presumption that we can learn what we ought to do, or are compelled to do, by studying other species. For Huxley, moral progress depends on rejecting nature as a model. Knowledge of evolution, he says, may explain how the good and the evil in humanity have arisen; but it cannot tell us anything new about reasons for preferring the good.

Huxley, therefore, was concerned not only with science and logic but also with their moral implications. He was uninhibitedly rude about those who extol war and justify the 'extirpation of the weak, the unfortunate, and the superfluous', on the ground that it is permitted by natural law. Such people, he said, 'count the physician as a mischievous preserver of the unfit'. They marry on the principle of the stud and their 'whole lives, therefore, are an education in the noble art of suppressing natural affection and sympathy'. Nobody writes like that today, which is perhaps a pity.

Huxley was a Darwinian who acknowledged the limitations of social Darwinism. Another, A.R. Wallace, held his own species to be unique in its possession of a moral sense: for us, he said, the great law of natural selection is partly neutralised; hospitality, he held, is a general human virtue opposed to the laws of nature. He might have added the custom of adopting unrelated children as an extreme and widespread kind of hospitality.

Other opposition, less well argued, came from politics. P.A. Kropotkin (1842–1921), a leader of the anarchist movement (and of peace biology), was convinced that natural selection produced, in each animal species, not belligerence but instincts of cooperation: these, he said, aided the struggle against enemies and against other environmental hazards. In his *Mutual Aid* (1902), he attributed human altruism and other social impulses to our evolutionary past. Unlike Bagehot, Haeckel and many others, Kropotkin thought highly of tribal groups, such as the African 'bush' people, Australian Aboriginals and the Inuit (Eskimo). They were, he said, without the materialism and greed prominent in civilized society.

CLINICAL DARWINISM

Despite these objections, both logical and moral, Bagehot's lead was energetically followed. A famous exponent was an English neurosurgeon, Wilfred Trotter (1872–1939). His *Instincts of the Herd in Peace and War*, first published in 1916, was many times reprinted and is still cited. Like other such writings, it combines authentic concern about human violence with absurdities.

Psychology, Trotter believes, if combined with other branches of biology, can guide human action and can even predict human behaviour. So he turns, like Aesop writing his fables, to the animal kingdom, where the herd instinct, he says, takes three forms: the aggressive (that of the wolf); the protective (the sheep); and the socialized (the bee). Modern civilized people emulate the bee. Germans, he writes, are not civilized but lupine, that is, aggressive. (This was during the first world war, when his countrymen were being urged to hate Germans— 'the Hun'.)

So Trotter combined an analysis of human society, based on animal stories, with crude wartime propaganda. He was therefore typical of those (still with us) who use animal exemplars: he chose species which suited his argument without regard to their actual behaviour; and he ignored the quite different conduct of other species. He was also typical in another way. Although humanity had evolved by natural selection, and was therefore, it seemed, riddled with destructive, animal instincts, Trotter hoped that 'the conscious and instructed intellect' could bring about a moral advance.

In Trotter's time and later, Freud's psychoanalysis provided a subspecies of *Homo pugnax* which became immensely influential. In the late nineteenth century, psychiatry had adopted ideas derived from biology. A human being was regarded as an organism driven by instincts of survival and breeding. Correspondingly, Freud was influenced by Darwinism. He held, like

Trotter, that 'the fateful question' for humanity is whether 'the human instinct of aggression and self-destruction' will be mastered. He proposed an elaborate account of human motivation in which aggression is prominent. In *Civilization and Its Discontents* he writes of human beings as

> creatures among whose instinctual urges is to be reckoned a powerful share of aggressiveness . . . The inclination to aggression is an original, self-subsisting instinctual disposition in man, and . . . constitutes the greatest impediment to civilization.

Many people, physicians, therapists and others (including psychiatrists' patients), seem to have found Freud's ideas helpful. As Thomas Nagel has written, Freud taught us to look for concealed meanings in human conduct—the hidden, symbolic significance in our actions. He also encouraged an enhanced sensitivity to the erotic. His descriptions of individual patients resemble the writings of poets and novelists. His nearest approach to scientific psychology is in his classification of personalities. The obsessional person is excessively orderly, obstinate, stingy, often unfriendly and dependent on the performance of strict rituals. A reader who knows such a person may expect him or her to overreact on making a mistake and to be furious if shown to be in error. Another type is the extreme narcissist, whose self-admiration prevents concern for others and leads to the delusion that success in any enterprise is certain. Failure can lead to deep depression or uncontrolled rage.

Freud's explanations of the varieties of wrath and antisocial violence are, however, based not on accurate descriptions but on imaginary internal processes, called 'drives' (*Triebe*, often translated as instincts), of which the stored energies need to be released. The energies are inventions: they have no physical existence. (Compare chapter 1 on fantasies about repression. More on drives later.)

During the rest of the twentieth century writers, who in effect—since they are human—call themselves *Homo pugnax*, have continued to influence opinion without exerting themselves in relevant research or critical study. Slogans such as 'the territorial imperative' have appeared in the media. Here is an

'Aggression' dissected	
Individual violence	**Group violence**
Assassination for pay	Football hooliganism
Assassination for a cause	Riot
Murder or assault by an insane person	Police action against protesters
Hanging a hated person during a race riot	Revolution
Rape	Counter revolution
Wife-beating	Genocide
Road rage	Ethnic cleansing
Thrashing a child	Persecution in the name of religion
A child's temper tantrum	Annexation
Bullying	War

incomplete list of human activities which have been put under the catchall heading of 'aggression'. If items from animal behaviour were added, the list would be very long: it would include, for instance, *defence* of territory, even by bird song. Each item requires separate study. Some of us therefore urge banishment of 'aggression' and related terms from serious writing. As a result we sometimes provoke animosity, though not so far, in my experience, actual assault.

PURPLE PATCHES

One reason for the popularity of *H. pugnax* is the entertaining character of the stories about him. Here are specimens from works which have been widely approved by reviewers and the public.

First prize should go to Raymond Dart, an Australian anatomist who described many important fossils of 'man-apes' (*Australopithecus*) in South Africa. For several million years, before our own genus, *Homo*, appeared, these enigmatic Primates sparsely occupied large regions of Africa. They seem to have walked much as we do but their brains were ape-sized. Dart was convinced that he understood not only their skeletons but also their social lives. He surpassed Kipling when he wrote that they were

> carnivorous creatures, that seized living quarries by violence, battered them to death, tore apart their broken bodies, dismembered them limb from limb, slaking their ravenous thirst with the hot blood of victims and greedily devouring livid writhing flesh.

Their quarries, according to Dart, included their own kind, for they were cannibals.

K.Z. Lorenz (1903–1989), a successful populariser and, like Freud, an Austrian physician, took up this theme.

> There is evidence that the first inventors of pebble tools, the African Australopithecines, promptly used their new weapon to kill not only game, but fellow members of their species as well.

And, on later fossil forms, he writes: 'Peking Man, the Prometheus who learned to preserve fire, used it to roast his brothers.' Unlike Dart, Lorenz published little original research. In particular, the study of our possible ancestors he left to scientists.

Much has been learned by years of patient, microscopic scrutiny of fossilised teeth and by comparisons with those of living forms. The resulting picture is of eaters of fruit and seeds. Some australopithecine skeletons, found in caves, seem to have been gnawed by members of the cat family. Our man-ape ancestors (if they were our ancestors) therefore appear as vegetarians preyed upon by carnivores, such as leopards. The idea that they or their successors were cannibals is a fantasy.

An ungentle reader who finds this a disappointing anticlimax may turn to Robert Ardrey (1908–1980), a playwright to whom we owe the motto, 'the territorial imperative'. Ardrey's dramatic portrait of humanity is of a Primate incessantly fighting for ground. The passage below is from his book, widely sold and discussed, which he calls 'a personal investigation into the contemporary, little-known accomplishments of the natural sciences'.

> A wonder of nature, mystifying and beyond all easy answer, is that the biological nation immediately appeared when true lemurs emerged from the long mammalian night. Could we only know better the animal psyche, we might find that terror of the day and subconscious remembrance of the monster combined to command the most perfect of primate defensive weapons though real need was lacking in a time before leopards were born. Could such a thesis be a subject for demonstration, we might know ourselves better.

Or we might not.

MEN OF PROPERTY

Ardrey equated animal territories with human property and defence of local or national boundaries. Others, ranging from psychotherapists to zoologists, have done the same. Land tenure has been represented as a fixed human instinct. The popularity of this notion is very strange. Not all animals, even mammals, are territorial. Among those that are, each species has its own kind of territory: it may be no more than a nest and its immediate surroundings, held only in the breeding season, for instance, that of a wren (*Troglodytes troglodytes*) which occupies a few cubic metres. Or it may be permanent and cover many hectares, such as that of a single tiger (*Panthera tigris*) in an Indian jungle or a herd of vicuña (*Vicugna vicugna*) in the Andes. Among our nearer relatives, many groundliving monkeys seem not to be territorial at all. These are facts.

Other facts are found in an obvious, contrasting feature of human landholding (and lack of it): its immense variety within our one species. It includes:

- nomadism without landholding;
- communal occupation of cultivated land;
- feudal ownership by lords who have both rights and duties;
- private occupation and production for the market;
- buying and selling land;
- municipal or national ownership.

The concept of property in land, as something which can be sold or given away, is a modern one: it is not 'instinctive'. For about six centuries, in the European Middle Ages, each village typically had common ground managed by a village council. In the sixteenth century, the commons began to be enclosed for private ownership, often to produce wool for the market. The livelihood of uncounted families disappeared. Hence followed a bitter saying, that sheep eat men. (For more, see Jeremy Rifkin's *Biosphere Politics*.)

I now give a case history to illustrate further our multifarious attitudes to land. When British and other Europeans colonised Australia, large 'enclosures' led to conflict. The newcomers treated Australia as unoccupied land (*terra nullius*) and converted great areas to stockbreeding and growing crops. But much of the continent was occupied by Aborigines. They had elaborate languages and complex customs and religious beliefs; and they were equipped with highly developed skills which made possible their survival, without agriculture, in harsh environments. They did not own land, still less buy or sell it. They had instead policies for their environment related to acquiring food and water and to sacred practices. Youths were taught the features of the region occupied by the group, and they learned obligations to the group and to the natural world.

When the northerners arrived, they wanted sole rights in cultivated ground. Correspondingly, early in the nineteenth century, some described the Aborigines as degenerate, brutal cannibals, unworthy of a white man's consideration. Others disagreed: the native inhabitants of the land, they said, were peaceful and had a 'plain and sacred right' to their own soil; but this opinion was, for the most part, disregarded.

The Aboriginals survived both white hostility and the accompanying new diseases. Yet, in the 1990s, the debate has been renewed. Even the persistent libel of cannibalism has reappeared. A decision in the Australian High Court of Justice, in favour of aboriginal land rights, has been received with outrage by a number of Australians, some of whom hold pastoral leases on the land. The lease holders have been represented as struggling farmers threatened by aboriginal access to sacred sites. Supporters of aboriginal rights reply that such access, peacefully negotiated, is harmless and that much of the land is in any case owned or occupied by wealthy persons or consortia, many outside Australia. Some resident pastoralists have therefore formed a society, Rural Landholders for Coexistence.

This digression is not to advertise the plight either of Aboriginals or of pastoralists (though it will no doubt be clear where I stand on this). The point is that we have here an example of human occupation of land which raises questions concerning religious practices, morals, legal principles, human violence, economics, politics, historical change and attitudes to all these.

Similar questions arise from the property relationships of other complex societies. The diversity of these relationships has no counterpart in the social interactions of any animal species. To equate them with animal territories is a ludicrous misuse of biological science.

DOTTY NATURALISM: *HOMO MENDAX*

An extreme case of likening people to animals is the portrait of humanity as the lying species. Many caterpillars are food for keen-eyed predators, such as birds, but are marvellously concealed (cryptic colouring). Other insects are brilliantly coloured but are poisonous or taste horrible (warning appearance or aposematism): when a bird has tried one, it avoids others like it. But some insects *resemble* an aposematic species but are not themselves toxic or distasteful (mimicry): we ourselves sometimes flinch from an insect, striped in black and yellow like a wasp, even if it is only a harmless beetle.

Accordingly, lying and other forms of deceit have been said to reflect our animal ancestry and to be instinctive features of human nature. We become *Homo mendax*.

The *deceptive* appearances of animals are an outcome of a long, slow evolutionary process. Ancestral beetles did not say, 'Aha! we'll grow orange stripes, pretend to be wasps and baffle the birds!' And human beings rarely go about either camouflaged or with a terrifying appearance. We do sometimes resort to *deceitful* (or mendacious) conduct; but this, though it may be morally wrong, is often a product of reason and calculated to give the deceiver some personal advantage.

If a rational and reliable account of lying is required, it is necessary to study not caterpillars or beetles but human beings. Most people, when they tell lies, or try to lie, are found to have difficulty in doing so convincingly. (Recall the work of Ekman, page 14.)

The general interest of this egregious fantasy is that it has been taken seriously by social scientists and even by some biologists. It is another case of blundering biological naturalism and of the collapse of the critical faculty found so often in human biology.

COMPARISONS ARE NOT NECESSARILY ODIOUS

The errors of naturalism do not oblige us to discard comparisons of *Homo sapiens* with other species. Features which we share with animals are easily found and can lead to useful conclusions. Most of our knowledge of human physiology depends on the study of animals. Late in the nineteenth century, a laboratory experiment helped to found the science of nutrition. Mice were fed on a purified diet of all the substances then known to be essential: carbohydrates, fats, proteins and inorganic salts. They died. Later, others were given a similar diet but with milk added. They survived. The milk, as we now know, contained small amounts of essential substances, the vitamins.

One vitamin is ascorbic acid (vitamin C). Without it, a human being eventually develops scurvy—a potentially fatal condition once familiar to sailors fed largely on salt pork during long voyages. I choose vitamin C as an example, because it carries a warning. Most mammals, including mice, rats and rabbits, can do without it in their diet: they synthesise it from other substances in food. In this respect, these commonly used laboratory animals are not models for humanity. The guinea-pig (*Cavia cobaya*), however, can develop scurvy. It can therefore be used in research on the effects of dietary vitamin C. But even here a problem remains: guinea-pig scurvy is not quite the same as ours; it appears much more quickly.

The case of ascorbic acid therefore shows both the value and the hazards of using other species to explain human physiology. Although we, *Homo sapiens*, have much in common with other mammals, in no case may we *assume* a likeness. This, as we now see once again, applies with special force to our social lives.

DEMONIC MALES AND GAY BONOBOS

Today, the myth of *Homo pugnax* survives in spite of its often exposed absurdities. As an example, I choose—rather unkindly—a book, published in 1996, by two Americans—an anthropologist, Richard Wrangham, and a lecturer in English, Dale Peterson—entitled *Demonic Males: Apes and the Origins of Human Violence*. On the jacket, leading anthropologists and primatologists tell us that the book explains why war and rape are the natural inheritance of the 'human ape'; that we share with chimpanzees the systematic killing of neighbouring males by organized war parties (not observed in my suburb); and that the book is essential reading for anyone who wishes to understand human violence.

But the experts either had not read the book or, at least, had not understood it: much of it is about the bonobo or 'pigmy chimpanzee' (*Pan paniscus*). Among bonobos, we learn, the sexes are equal. Sometimes the females seem to dominate: at meals, they eat first. Before a meal, all have sex with each other, commonly between members of the same sex: males rub penises; females indulge in mutual clitoral stimulation. Female bonding is prominent: females often form powerful groups.

The book is presented as a product of science; but its basis is again biological naturalism. Much is in the tradition of the bestiaries of the Middle Ages. Their objective was to promote moral conduct. In one, edited by T.H. White, the twelfth century author remarks how like we are to monkeys; but he seems to find monkeys rather disgusting and prefers lions.

> The compassion of lions is clear from innumerable examples—for they spare the prostrate and they allow such captives as they come across to go back to their own country . . . The nature of lions is that they do not get angry unless they are wounded. Any decent human being ought to pay attention to this: for men do get angry when they are not wounded . . .

So for him a remedy for our antisocial violence is to emulate the lion.

Wrangham and Peterson too are worried by human violence. They have, they say, 'a new view of male chimpanzees [not bonobos] as defenders of a group territory, a gang committed to the ethnic purity of their own set'. Here they write of chimpanzees as if they were human, and imply that they can have a concept of 'ethnic purity'. They also ask dramatically: does chimpanzee behaviour imply that human killing is rooted in prehuman history? Their answer seems to be, 'Well, yes and no'. Because, later, we learnt that the neighbouring species offers a way out.

> Bonobos have shown us that the trap can be broken through female alliances. Among humans, a *direct equivalent* would be if women always remained together, day and night, in groups so large and well armed that they could always suppress the hostility of rowdy, aggressive men. [Emphasis added.]

This is not, I think, intended as a joke, though the authors do state that the idea is fantastic. They also say that 'with an evolutionary perspective we can firmly reject the pessimists'. By the pessimists, they evidently mean those who say that we are biologically forced to be violent.

Fortunately, humans can create other possibilities. Male demonism is not inevitable. Its expression has evolved in other animals, it varies across human societies, and it has changed in history.

Hence the 'evolutionary perspective' is no more than a recognition of diversity. Evolution has produced very different kinds of society in closely related species: chimpanzees and bonobos are genetically very close and socially very different. Their cousin, the gorilla, is different again. And, more important, as the authors say, violence and male conduct vary greatly within their own species, *Homo sapiens*. How all this variety evolved is not known.

In the tradition of earlier writers, on this shaky ground the authors propose political action.

In true institutional democracies, political power ultimately comes from the ballot box. And it is to the ballot box that women in the real world can mass themselves . . . and break through the trap defined by male interest.

So, in exact opposition to some others, notably the famous E.O. Wilson, who also write at length about the 'human animal', the authors urge that women should be more influential in public affairs. Women vigilantes are to band together like female bonobos. They are to vote together—I am not sure for exactly what; and so they will defeat the demonic males of whom, presumably, the authors and I are examples.

But they do not say that we should copy all bonobo social behaviour—certainly not at meals. And as for the belligerent, male chimpanzee—he is just as close to us biologically, but he will not do at all. So, again like other writers, they choose a species which suits their doctrine, and select aspects of that species' behaviour which conveniently match their teaching. In this way any form of social structure can be represented as natural or instinctive for humanity: hunting or vegetarianism; bellicosity or pacifism; patriarchy or matriarchy; the extended family, the nuclear family or a solitary existence. Some readers may applaud the support of these authors for women's liberation, others may oppose it; but neither position should be propped up by animal stories.

THE DEVELOPMENT OF VIOLENCE

Writings of this sort are more than amusing aberrations, for they have been widely noticed in the media and have helped to influence opinion on matters of life and death. Social surveys in eighteen countries have found 40 or 50 per cent of people to accept war and other forms of violence as part of the nature of humanity and fixed in our genes by a vague 'Darwinian' process.

If the picture of humanity as *H. pugnax* were true, attempts to curb the many kinds of antisocial violence would presumably be fruitless; but facts and serious research tell a different story. In 1998, figures were published on the annual number of murders to each 100 000 inhabitants for all European and American cities with a population of above a million. Washington, DC, with 69.3, scored highest, then followed Philadelphia, Dallas, Los Angeles and Moscow. At the bottom of the list were London (2.1), Dublin, Rome, Athens and Brussels (0.4). Of the many factors which influence these barely credible differences, some are known or can be surmised. In cities with many murders, the problem is to alter the conditions which cause them. In this endeavour, talk of an instinct, a drive or genes for violence is irresponsible.

A crucial question is: how do violent and pacific attitudes develop, especially in early life? In experiments, children have been shown people bashing a large doll. Other children ('controls') saw the same people and objects in peaceful situations. Afterwards, the children tended to imitate what they had seen. Correspondingly, if destructive acts are violently punished, the outcome is often imitation of the punishment, in the form of worse violence by the offender.

One proposal has been based on the idea of catharsis or release of 'tension'. Hostile and destructive behaviour has been attributed to pent up 'aggression' which needs to be expressed. (Compare the Freudian drives mentioned on page 45.) But, when thoroughly tested, rewarding children for expressing violence has either had little effect or has enhanced the unwanted behaviour. Again, much depends on what older people say and do, especially parents and others whom a child can adopt as models.

The idea of catharsis has also been tested on adults. Volunteers have been subjected either to insult or to pain.

Both are liable to induce a rise in blood pressure and heart rate. Some subjects responded angrily but others reacted in a friendly way. (Perhaps they were familiar with the proverb, 'A soft answer turneth away wrath'.) Both kinds of response were accompanied by a lowering of blood pressure and heart rate. In other experiments, subjects were trained to respond to insult by giving themselves shocks, that is, to behave masochistically. This too relieved 'tension'. Such findings, based on well designed scientific enquiry, illustrate not a fixed, 'aggressive' response to provocation, but human variation. The conduct of the subjects depended on personality, including attitudes socially acquired.

The most obvious source of differences is gender. Kirsti Lagerspetz and her colleagues have recorded the expression of anger among children of eleven to twelve years. The behaviour studied ranged from kicking and hitting to telling tales, and from abusive language to telling teacher. The principal findings concerned differences between the sexes: girls expressed anger slightly less than boys but made more use of indirect forms of anger. These findings, from the Finnish city of Turku, illustrate the diversity of the actions called 'aggressive'. Moreover, no guarantee exists of similar results from identical researches in Tokyo or Timbuktu.

Just as children learn skills by imitating older people, so they can learn violent habits from what they are daily accustomed to see and hear. During the second half of the twentieth century, in many countries both violent crime and also the amount of violence presented on television have greatly increased. L.R. Huesmann and others have reviewed half a century of research which has revealed a consistently ill effect of exposing young children to such entertainment: it has shown how urgent is reduction of the violence which has become conventional in popular fiction. Children are not forced to be belligerent by a genetically fixed 'animal nature'. They grow up in distinctive environments and interact with them. During their development, they can be encouraged to be kind or unkind, friendly or hostile.

As we know, in the light of rigorous research the melodrama of cavemen struggling with other cavemen (and with or for their wives?) disappears. (Nor should the neglected wives and children be suspected of inherent bellicosity.) Similarly, serious study relieves modern humanity of the burden of an untreatable madness. Science therefore speaks to us as rational beings. Unlike the stories about *Homo pugnax*, it provides guidance for action. It presents a need for restraint in our social lives and in our preferred fiction and for friendly and pacific attitudes to those around us.

WAR

In that case, a reader may ask, why do people go to war? During much of their existence, human groups have lived in peace. Even now, we are not all subjected to incessant conflict. (Look around.)

Warfare on a large scale seems to have been a recent invention—a product of the gathering of people in cities, that is, of civilization. The earliest traces of defensive fortifications are dated about 5500BC. In Eastern Europe, they are as late as 3500 BC. 'War', however, does not refer to a single kind of event. Here are two examples of activities, both of which entail group hostility.

First, in the Highlands of New Guinea, boys are systematically trained in violence; and, as men, they take part in regular battles. They don elaborate costumes, take bows or spears, shout challenges and, when the other side is ready, they advance and clash. The battles end in time for everyone to get home before dark. The victors gain no booty or ground—not even women. In this case, all the people concerned run into some danger. The nearest counterpart in the reader's daily experience is, perhaps, a football match, but the conduct of the people in New Guinea is a little more dangerous. It is also characteristic of a single tribal group. It is no more 'species-typical' of *Homo sapiens* than are American or Rugby football.

The other example is the Gulf War of 1991 (fully discussed by Hamid Mowlana and others). That event can hardly be attributed to anybody's biological compulsion: the American president, George Bush, did not storm into the bunker of his leading opponent, the Iraqi president, Saddam Hussein, and offer him choice of weapons. War had been decided on, some months before it began, after prolonged discussion. This was a clandestine decision, made by bankers, lawyers and generals on

the basis of calculation. It was, in a famous phrase, a 'continu-
ation of policy by other means'; and not everyone, even in US
ruling circles, supported it. Historians have shown the origins
of major wars to resemble that of the Gulf War: typically, they
are embarked upon after debate, often secret, among a nation's
rulers.

The Gulf War, though unusual in being onesided, resembled
other wars in the public response to it. 'War fever' was worked
up after the decision for war was made. Accepting mass violence
did not come easily to Americans, any more than it does to
anyone else. George Ball, a former American Under-Secretary
of State, has described how, in the USA, it was necessary for
President Bush to use vituperative language and hyperbolic
denunciations of Hussein in order to 'get the people to fight'.
This propaganda was successful. Large scale indiscriminate killing became 'acceptable'. (Compare page 44 on the hatred
encouraged during the first world war.)

An opposite aspect is shown by some events in Iraq. The
men of British tank regiments regularly allowed Iraqi crews time
to get away, before their tanks were shot up. And some of the
Iraqis refused to fight: they either surrendered or retreated as
quickly as possible. British troops were also described as being
upset by the wretched state of many of the Iraqi prisoners of
war. These items, from reporters on the spot, had little pub-
licity. They can be paralleled from other wars. Yet, unlike
violence, the 'quality of mercy' and an impulse of friendship are
rarely proposed as part of human nature.

Here is a relevant passage, slightly altered from my *Biology
and Freedom* , on the early stages of trench warfare during the
first world war.

> In 1914, during what a historian, John Buchan, called 'the
> extraordinary truce of Christmas day', men of the opposed
> armies met in joint celebrations. In one place, 'a famous clown
> in the German trenches occasionally went through perfor-
> mances amid the applause of both sides'. In others the truce
> included games of football. As Buchan remarks, 'outposts have
> always fraternized to some extent—they did it in the Peninsula
> and in the Crimea'. In committing these gross breaches of
> discipline, men succumbed to impulses of friendship and tol-
> erance which are at least as much features of humanity as are

violence and hatred. No doubt, by 1918, nearly all were dead or seriously injured. Yet their most significant contribution to history was not military: it was the brief, peaceful interlude in which they took part—provided that we remember it.

HOMO EGOISTICUS

Although most of us are peaceable most of the time, almost anybody can be provoked to violence. That truism apart, the concept of *Homo pugnax* conveys nothing useful. After 1960, however, another impostor, *Homo egoisticus*, emerged with a seemingly more impressive evolutionary background. Its begetters, like their predecessors, are preoccupied with human depravity, but their range extends beyond human violence.

The leading sin is now selfishness. It used to be thought, even by biologists, that animal action is often directed to the survival of the species and is therefore 'unselfish' or 'altruistic'. A bee, perhaps, loses its life by stinging an intruder in defence of the hive; or the leading male of a herd of antelopes or of a troop of monkeys succumbs to lionesses and allows the others to escape.

These are descriptions of animal conduct in human terms: loyalty to a group or a cause, hence self sacrifice, are features of human social life. But evolutionary theory does not allow such attitudes among animals. An animal which handicapped itself on behalf of others would be less fit, in the Darwinian sense, than others whose behaviour preserved only themselves; for, if its difference from rivals, in this respect, were genetically determined, the chances of survival of its descendants would be diminished. We should therefore expect an animal never to harm itself by helping others. That is, it should not behave altruistically. (I am using 'altruism' here with its primary meaning: *concern for others as a principle of action*.)

At first sight, the only exception would be conduct which helped the animal's offspring. This would contribute to the survival of the animal's genes and would therefore be compatible with natural selection. Theory, however, admits two more exceptions. First is behaviour called 'kin selection', which helps close relatives: these have nearly all the same genes as the helper. Second is behaviour which helps another of the same species, when the other may be expected to be helpful later ('reciprocation').

The modern 'Darwinian' picture is therefore a system in which some types are automatically more successful than others. The relevant differences between types are genetically determined. Success is measured, when (rarely) it can be measured, by the production of descendants with the same genes.

This amoral concept has been applied to the human species. A leader of the movement, the American entomologist E.O. Wilson, in his much discussed *Sociobiology*, proposes a reduction of the hum- anities and social sciences to evolutionary biology. He writes that 'an organism is only DNA's way of making more DNA'. Accordingly, in the human brain, 'the hypothalamus and limbic system are engineered to perpetuate DNA'; and Wilson attributes 'hate, love, guilt, fear' to processes in 'emotional control centers'. These hypothetical structures, reminiscent of the bumps of phrenologists, 'evolved by natural selection. That simple biological statement must be pursued to explain ethics and ethical philosophers . . . at all depths'. In a later work, *On Human Nature*, he writes:

> Human behaviour—like the deepest capacities for emotional response which drive and guide it—is the circuitous technique by which human genetic material has been and will be kept intact. Morality has no other demonstrable function.

Wilson's main message, as it has reached a large public, is this:

> The question of interest is no longer whether human social behavior is genetically determined, it is to what extent. The accumulated evidence for a large hereditary component is more detailed and compelling than most persons, including even geneticists, realize. I will go further: it is already decisive.

This dogma is presented as a product of biological science. The confident, magisterial tone is, however, not supported by cogent evidence or logical argument. On the contrary, in the same book, we find self contradiction:

> . . . let me grant at once that the form and intensity of altruistic acts are to a large extent culturally determined. Human social evolution is obviously more cultural than genetic. . . . The sociobiological hypothesis does not therefore account for differences among societies.

Such incoherence runs through all human sociobiology.

A narrow focus on DNA has also led some writers to allot genes to aggressiveness, xenophobia, cheating, lechery, lying, nepotism, spite and other evils, all of which are attributed to the past action of natural selection. These genes are inventions: nothing of the sort exists in genuine human genetics. They sound like an attempt to find a biological basis for the seven deadly sins of the Christian Middle Ages: pride, lust, anger, greed, envy, sloth and covetousness.

Wilson himself, however, now seems to have discarded the notion, implicit in his best known writings, that human beings are puppets of their DNA. He has, most commendably, turned to the formidable problems of preserving the biosphere. In a recent work, *The Diversity of Life*, he writes:

> The stewardship of the environment is a domain . . . where all reflective persons can surely find common ground. For what, in the final analysis, is morality but the command of conscience seasoned by a rational examination of consequences? . . . An enduring environmental ethic will aim to preserve not only the health and freedom of our species, but access to the world in which the human spirit was born.

Here he gives up the amoral *H. egoisticus* mindlessly and 'selfishly' striving for the survival of a set of genes. Instead he accepts his and our freedom to make rational and moral choices.

HOMO LIBIDINOSUS IN ITS NEST

Not all Wilson's disciples have followed this example. In *On Human Nature* Wilson had written:

> It pays males to be hasty, fickle and undiscriminating. [Is he describing himself?] In theory it is more profitable for females to be coy, to hold back until they can identify the male with the best genes . . . Human beings obey this biological principle faithfully.

This is not a scientific finding on humanity. It is a statement of an attitude to the social roles of the sexes. To match it, a new school of sociobiologists have spawned a mutant of *H. egoisticus*. A journalist, Matt Ridley, tells us that the problem of identifying a true human nature has been solved. His book, supported by another populariser, Richard Dawkins, is described on the cover as a 'brilliant examination of the scientific debates on . . . sex and evolution': men want 'power and money in order to pair-bond with women who are blonde, youthful and narrow-waisted'. This vulgar twaddle has no biological basis; but the book itself does offer some light relief.

Ridley's account is an extreme example of the error of naturalism, for he likens us not to one species, such as the bonobo, but to many. Some of us are monogamous and so resemble albatrosses (*Diomedea exulans*). 'Every female gets a model husband; they share equally the chores of raising the chick.' But alas! others are polygamous and resemble elephant seals (*Mirounga leonina*) with 'the males battling and exhausting themselves, and dying often in the vain attempt to be the senior bull'. Evidently, when we are in this guise *Homo pugnax* is among our near relatives. Other species appear when Ridley writes: 'It is my contention that man is just like an ibis or a swallow.'

The title of Ridley's book is *The Red Queen*—after a character, from Lewis Carroll's *Through the Looking Glass*, who was notorious for running very hard without getting anywhere. So is the author satirising sociobiology? On the evolution of our intelligence Ridley writes: 'animals use communication principally to manipulate each other.' 'To manipulate' means 'to manage by unfair means or contrivance' and this goes with a misanthropic picture of humanity. Finally, Gall and his phrenology are outdone: Gall held the cerebellum to be for sex, but from *The Red Queen* we learn that the cortex is for courtship. No valid evidence is given. Hence the book, though it has been taken seriously, emerges as an elaborate practical joke: in fact, a mare's nest.

Another populariser, an English zoologist, Robin Baker, has combined sex with aggro in a story about 'sperm wars'. He copies earlier ultradarwinian writers by saying that the sexual behaviour of a woman is moved by her need to maximise her Darwinian fitness and so to have children by several promising fathers. To this he adds the insult that, at a certain stage in the monthly cycle, women are liable to be compulsively unfaithful to their husbands.

The idea of sperm wars is a recent example of the ancient practice of treating women as irrational beings below the level of men. Two and a half millennia ago, the Greek founders of

Western political science and philosophy regarded women as degenerate males. Since then, writings by men about women have continued, in effect, to assure their male readers of the inferior quality, as human beings, of their mothers, wives, mistresses, daughters and others.

After 1859, these romancers—like those who advertise *Homo pugnax*—have been able to invoke 'Darwinism'. In doing so, they distort or invent scientific findings. They also disregard an everyday fact: many women now decide, after careful thought and—sometimes—discussion with a partner, just when they will have a child, and by whom. Their decisions make use of new knowledge and are based on reason.

The caricatures of the human species, described in this chapter, when they have a serious meaning, imply that to explain human social life all that we need is a 'biological' account of humanity. Such notions have been repeatedly refuted—recently, for example, in the excellent *Myths of Gender* (1985) by Anne Fausto-Sterling, in *Vaulting Ambition* (1985) by Philip Kitcher, in my *Biology and Freedom* (1988) and in *Biology as Ideology* (1991) by R.C. Lewontin. Yet these fantasies, or others like them, continue to be published and held out as science. Human social conduct is presented as forced on us by the past action of natural selection and dominated by instinctive violence, egoism or sex: the possibility of intelligent action, or of disinterested behaviour (that is, of altruism in its primary sense), is played down or treated as an illusion. So is friendship.

It is strange that writers, themselves personally agreeable and privately virtuous, should energetically promote such libels on their own species; and it is still more strange that their logical and biological errors should be taken seriously. As we see in later chapters, their portrayal of the human condition ignores humanity's most prominent features.

CHAPTER 4

INTERMEZZO ON INSTINCT

If music be the food of love, play on;
Give me excess of it.

SHAKESPEARE, *Twelfth Night*

IN ATTEMPTS TO EXPLAIN them, aggression and owning property, like other human propensities, have been called instincts; but that is only naming. People who resort to such purely verbal explanations are like Molière's notorious (and successful) doctoral candidate, who accounted for the action of a soporific drug by saying that it had a 'dormitive essence'.

In everyday speech, 'instinct' commonly refers to intuition or unthinking skills, acquired by experience. In his novel, *Kangaroo*, D.H. Lawrence (1885–1930) describes a character as having 'that alert instinct of the common people, the instinctive knowledge of what his neighbour was wanting and thinking'. Such abilities are indispensable for our social lives but they are hardly relevant to instinctivist, supposedly scientific, accounts of human conduct.

In popular science, instinct is often a mendicant word—one that begs for a meaning. Here is E.O. Wilson again, in *On Human Nature*.

> . . . innate censors and motivators exist in the brain that deeply and unconsciously affect our ethical premises; from these roots, morality evolved as instinct.

The innate censors are an invention: neurophysiology knows nothing of them; the roots are a metaphor which hardly helps understanding; and the last four words quoted, if they have a meaning, imply that our moral principles are fixed, like the form of a spider's web or the route of a migrating bird. This is obviously wrong: moral principles vary greatly in different communities and at different times: they can be rationally debated and, as a result, altered.

THE SCIENCE OF 'INSTINCT'

In the authentic science of animal behaviour (ethology), 'instinct' has had two distinct uses. One refers to observable behaviour, the other, to internal causes.

Certain actions performed by animals look skilful and seem to develop without practice or the use of intelligence. They are common to a whole animal species and evidently adapt it to a particular way of living. They include the consumption of food, the building of nests or webs and concealing behaviour. They are therefore assumed to be direct products of natural selection.

Species-typical features of social behaviour include the signals between parents and young and between mates, and those that warn companions of danger. Counterparts exist in human action. A human infant smiles at the sight of a face and cries in response to pain or discomfort. These universal features of humanity match gestures and sounds made by the young of other species. Some facial expressions of adults too seem to be typical of *Homo sapiens*. Experimenters have used photographs or actors to represent emotions: expressions which reflect happiness, anger and gloom are then usually recognized throughout the world.

Mammals communicate with each other largely by means of odorous substances (pheromones) detected by a structure, the vomeronasal organ, close to the nasal cavity. Pheromones attract mates and have other functions. The vomeronasal organ is present in human beings and recent experiments have revealed effects of human pheromones. Young celibate women in college dormitories often have synchronized menstrual cycles. The synchrony is due to skin secretions. In addition, a mother can distinguish the odour of her infant on a garment from that of other infants of the same age. Children, too, prefer the odour of their mother to those of others. Yet more possible phero-monal influences are being studied. An Israeli psychologist, Aron Weller, suggests that they can eventually contribute to new methods of contraception.

These findings belong with what Darwin called the 'indelible stamp of our lowly origin'. Equally significant, however, is an opposite phenomenon—our liking for abolishing or concealing body odours by washing or by applying deodorants or perfumes. Such customs vary greatly among different communities. In the European Middle Ages, baths were often communal and nakedness was not shameful. But, in the late Middle Ages, physicians, notably those who advised the royal houses of France and England, regarded bathing as unhealthy. And, in the sixteenth century, the Reformation led to yet another change of attitude: for a time, the churches deplored bathing as immoral. Then, in the nineteenth century, baths came back into fashion for reasons of hygiene; but they remained private. Today we have a partial return to earlier attitudes. These changes have not been driven by 'instinct' or genetics. The variation is cultural. To understand it, we have to study the influence of conventional beliefs.

The idea of instinct, however, becomes seriously misleading when it refers not to observed behaviour or secretions but to mysterious internal processes. Animals and people tend to achieve certain ends, often by variable means and in the face of difficulties. Hence follows the notion of being impelled toward an end by an impulse or drive: hunger drive impels eating and so on. Sometimes, each instinct is represented as a source of energy (compare Freud's *Triebe*, page 45). A zoologist, Colin Beer, has pointed out that, in such writings, the energy is not measured but is said to 'flow', or to be 'dammed up' or 'consumed' or 'to spark over'. As these metaphors show, it is not the energy of physics or indeed anything observable.

Even in the 1990s, it was possible to read in scholarly works expressions such as 'the call of instinct'. In science, however, internal instincts or drives are now being replaced by what we can measure. Much fluctuation in activity is related to maintaining a steady internal state (homeostasis). To study internal causes of behaviour, a set point or target value may be looked for, say, a particular blood sugar level or body temperature. The behaviour is then seen as part of a regulating system which prevents departure from the set point.

Consider eating. To say that we have an instinct to eat tells us only what we know already: that (illness and deliberate fasting excepted) we all energetically consume or try to consume food. We may usefully ask: what internal processes impel (and stop) this essential activity? What is the physiology of 'hunger drive'? A long answer would be possible. An empty stomach encourages eating; so do a decline in blood sugar, a low body temperature or depleted fat stores in adipose tissue. As well, we can learn to choose from among alternative foods, according to need: salt or vitamin deficiency can influence what we choose to eat. The obverse is rejecting food which has caused an upset tummy—a common experience.

But, for human activities, even those as essential as feeding, physiological analysis has limitations. At meal times, food may be eaten without any internal promptings: habit has an influence; so have anxiety, boredom and greed. Choice includes a powerful social (cultural) element: what for many are nutritious delicacies (snails, insect grubs, octopus, frogs . . .) for others are repellent. Many people find even ordinary meat unacceptable.

Similar variety is seen in maternal behaviour. Many mothers refuse to feed their infants at the breast: they resort to wet nurses or, today, to bottles containing milk substitute. Such varying conduct can again be understood only by studying the cultural influences which act on people as they grow up. For more on medieval and modern customs, and their ecological implications, see Jeremy Rifkin's excellent book, *Biosphere Politics*.

Here is an incomplete list of human activities which might be called instinctive.

acting	calculating
coitus	cooking
exploring and tinkering	keeping pets
kissing	making music
making tools	ornamenting the body
playing games	showing off
taking drugs	talking
teaching	washing
wearing clothes	worshipping

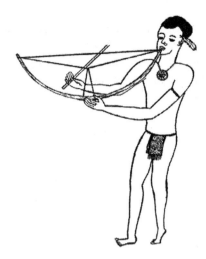

The universality of music. A music bow of West Africa. (From J.E. Lips, *The Origins of Things*)

To show further the limitations of 'instinct', I take one item which seems to occur in all human groups: making music.

THE 'FOOD OF LOVE'

Animal 'music' may be heard during courtship and mating and during defence of a territory. Sounds, such as bird song, also help in parental and filial conduct and in warning of danger. Although they sometimes depend on learning in early life, each song or sound is typical of the species (hence has been called instinctive); and each has a typical function which, with difficulty, can be identified by experiment.

Although music is typical of humanity, it is quite unlike what we hear from other species. People sometimes, like Orsino quoted at the chapter head, use music in courtship; but human mating does not crucially depend on it. People have used music during conflict, even some kinds of war; but military success can be achieved without it. Music is important in religious ritual; but religion can be practised silently. The uses of music are diverse, like the societies in which it is heard.

We are unlikely ever to know how song arose, perhaps in the Upper Paleolithic, but much is known of how musical ability develops in individuals. Infants of a few months pay attention to a sequence of sounds, such as a scale played on a piano: a change in the sequence is noticed and affects the heart rate. Infants also respond with their own sounds to the voices of adults and imitate them. Singing begins in the second year but is not yet very tuneful. The ability to recognize tunes and to sing them is, as a rule, fully achieved only by the eighth year. Children of that age can be taught to play the violin. Musical skills, however, vary greatly. Much remains to be learned about how they can be encouraged and taught.

The forms of music vary. The kind of music one enjoys usually depends on what is familiar: liking is culturally influenced. Perhaps the reader is familiar with the western kind, based on a chromatic scale with intervals of a semitone. With some modulation, it is written in a key which requires only seven notes. Other widely heard music, such as that of the sitar in India, uses intervals of less than a semitone.

Like our other social activities, music changes and develops. It has been described as a mathematical art which developed a special language of its own. A historian, A.W. Crosby, has called the written kind, which originated in the eleventh and twelfth centuries, Europe's first graph. The modern written form and orchestra finally emerged in the seventeenth century. Today we also have fascinating sounds composed with the aid of computers.

For those who make and listen to them, many of these various sounds have, like speech, a meaning: they can represent emotions, events such as storms and activities such as hammer on anvil. They also have what J.A. Sloboda calls 'embodied' meanings which depend on the relationships of the notes to each other: music is written in different keys, in major and minor, and at different tempi. To respond fully to these aspects of music, most people require much study.

Human variation appears in an extreme form in our responses to music: modern compositions, which many enjoy, send others up the wall; while a fugue by J.S. Bach, which some regard as a peak of musical achievement, to others is merely tedious. So music, though a universal source of joy, is immensely diverse in place and time; it has a complex social history, a similarly complex development in the individual and often an intricate structure. In all this elaboration, it is typical of other human skills and social practices. To call it an instinct would be to ignore nearly everything interesting about it.

THE TEACHING SPECIES: *HOMO DOCENS*

Musical skills are taught. A trite statement, the reader may think. Yet teaching is a central feature of human action. Is there then an 'instinct to teach'? This notion further exposes the limitations of 'instinct'.

The universality of teaching, and its fundamental social roles, are often overlooked. Instead, throughout history, we find mixed attitudes to teaching and teachers. The Talmud, the ancient Jewish system of laws, tells us that teachers should be an object of veneration. Yet in ancient Greece and Rome schoolmastering was the least esteemed and worst paid of the professions. The Greek word for education, παιδέ ια (paidéia), sometimes translated as 'culture', could also mean correction or punishment. Similarly, one of our own words for teaching, pedagogy, may signify only the imparting of knowledge; but, as the Oxford Dictionary points out, it has an implication which is 'usually hostile'.

The attitudes of teachers also vary. A pedagogical policy, once widely accepted but now given little support, is expressed in a notorious principle attributed to an English schoolmaster: that it doesn't matter what you teach boys, so long as they throughly dislike it.

In the previous chapter, the standardized ('instinctive') skills of web building spiders, migrating birds and others are contrasted with our own skills acquired, sometimes painfully, in response to schooling. The human ability to teach, combined with our capacity to learn from instruction, compensates for our lack of 'instinctive knowledge'. Teaching transmits information between generations. It makes possible rapid addition of novel practices to established traditions. We may therefore say that we 'inherit' knowledge from our ancestors. But the passage of information by teaching is quite unlike biological heredity. It is not a one-way process: information can be passed from children to parents and from children to other children. Most formal instruction is by unrelated persons; and it can cross national and racial boundaries.

In complex societies, teachers are specialists: they themselves have to be taught. What they transmit is diverse. Pupils may be told how to perform clearly defined operations, such as spelling 'pedagogy' or making an item of furniture with a prescribed structure; but they also learn open-ended skills, such as writing clearly and how to teach children. And pupils are taught *about* phenomena, such as those studied by historians, linguists or zoologists.

The forms and content of teaching reflect the variety of human cultures. A modern innovation, regarded today as essential in a democracy, is universal schooling, but just what should be taught in the schools is still debated. Now, however, that 'teacher' is no longer synonymous with 'torturer', one principle is coming to be generally accepted: that pedagogues should be free to encourage their pupils to reach conclusions after rational argument. For an effective democracy this too is crucial.

A human being can 'teach' some things merely by showing. The pupil then learns by imitation. But, in the most elaborate teaching, the instructor tries to persist until the pupil achieves a certain level of achievement or improvement. (I have called this 'teaching *sensu stricto*'—teaching in the narrow or strict sense.) The best results require empathy on the part of both instructor and pupil. (More in my review, published in 1994.)

With or without empathy, 'pedagogy' depends on the use of language. And language itself is, of all our peculiarities, the one that marks us off most clearly from other species. For the human species, *Homo loquens* is an alias still more appropriate than *Homo docens*.

The reader may ask what the last four paragraphs have to do with instinct. The short answer is: nothing at all. That is the point. Instinctivists try to explain human action, yet they fail even to describe what they are trying to explain. To say that the human species has an instinct for any of the activities in the list above implies that they may be regarded as distinct traits of behaviour driven by fixed, internal impulses. But in fact none forms a distinct unit and each has a history. In any generation, what is taught depends on decisions made at the time. It reflects what people *intend*. Biology does not authorize us to fall back on fixed instincts or genes and leave them to determine development. We have to make choices. This conclusion, as we see in the next chapters, matches the findings of modern biology.

CHAPTER 5

GENES AND CLONES

There was a young man who said, 'Damn!
How horrid to think that I am
    A system that moves
    In *genetical* grooves:
Not even a bus—but a tram!'

T HE BELIEF IN FIXED instincts, or in some other unavoidable destiny, has a long history. A classical scholar, E.R. Dodds, has written of the 'astral determinism' of the star worshippers of ancient Greece: behind its acceptance, he suggests, there lay, among other things, the fear of freedom—the unconscious flight from the heavy burden of individual choice which an open society lays upon its members.

Astral determinism or astrology, already encountered in chapter 1, makes us 'servile to all the skyey influences'. Modern genetical destiny is internal and is more like another ancient concept, that of an individual's daemon. F.M. Cornford (quoting from Plato, 428–347 BC), writes of

> the belief in hereditary guilt—those 'taints and troubles which, arising from some ancient wrath, existed in certain families', and were transmitted with the blood to the ruin of one descendant after another.

This credo, too, relieves one of responsibility. In the Middle Ages it was replaced by what Bertrand Russell called the 'ferocious doctrine' of original sin, which imposed guilt on those who accepted it. In his *Letters to the Romans*, Paul of Tarsus (died CE 65?), the principal originator of Christian doctrine, had asserted that all humanity is 'under sin':

71

There is none that doeth good, no, not one . . .
With their tongues they have used deceit;
The poison of asps is under their lips:
Whose mouth is full of cursing and bitterness . . .
Destruction and misery are in their ways:
And the way of peace have they not known.

According to another saint, the great theologian, Augustine of Hippo (354–430), when Adam and Eve ate the apple in the Garden of Eden they became corrupt and their descendants (that is, humanity) inherited their corruption. As a result, Augustine held all infants to be born damned; he accepted slavery as a form of judgement on the guilty; and most of us, slave or free, are destined for eternal punishment. The exact nature of the original sin seems never to have been decided; but this belief influenced Christian teaching for many centuries.

With the eventual rejection of original sin, Christians were no longer required to accept human distress with resignation. The social causes of misery could be sought and perhaps removed. So another, more acceptable burden was imposed: the need and the duty to take action against social ills.

Our next question—technically the most difficult in this book—is whether genetics after all provides a case for reverting to ancient fatalism.

GENES AS GENIES

To answer, we have to face again the unrelenting problems of heredity and environment. The subject has caused confusion especially since 1892, with the work of the German biologist, August Weismann (1834–1914). A century later, James Watson, famous as one of the discoverers of the structure of DNA, said: 'We used to think our fate was in the stars. Now we know, in large measure, our fate is in our genes.' Similarly, the editor of the leading American weekly, *Science*, once told his readers that 'in the warfare between nature and nurture, nature has clearly won'. (He does not explain his choice of 'warfare' as a metaphor; but by 'nature' he meant heredity—the genes; and, by 'nurture', the environment.) (For more, see *The Code of Codes*, edited by Kevles & Hood.)

The statements quoted are not even wrong: they are absurd.

They disregard a fundamental biological principle, conveniently illustrated by an incident in a macabre short story by Dorothy Sayers. A young man receives, as a legacy, a relative's stomach and intestines (preserved in formalin). So, in a perfectly proper and legal sense, he *inherits* his relative's guts. But the young man, of course, also has a complete alimentary tract of his own. His guts have not (in the same sense) been inherited from his parents or from anyone else: they have *developed*, with all his other organs, gradually, from a beginning in a fertilized egg.

Yet we still often say that our bodily structures and traits of character are inherited (or are 'genetic'). The two distinct uses of 'inherited' may be summed up in two sentences:

- *She inherited her mother's furniture, fortune and name.* This represents the legal or conventional meanings of 'inherited'.
- *She inherited her mother's looks and intelligence.* This statement belongs—if anywhere—to genetics or embryology. 'Inherited' is a metaphor. It refers to *resemblances* between parent and child which appear while the child grows up.

We are now concerned with the second use.

A human being (or any other organism) develops in a series of environments. The outcome, at each stage, depends on exceedingly complex interactions of gene products with impacts from outside. For a faithful account, we have to find out how genes work. Elementary texts are rarely helpful, for they make it easy to believe in genes as having simple actions. Even today, geneticists still often speak of a 'gene for' a trait. A famous biologist, Peter Medawar, once described this practice as slovenly.

Granted, occasionally it is not misleading. A reader who has the blood group O carries in each cell two copies of a 'gene for' a particular protein. If, instead, one of those two genes were slightly different, the reader could be of blood type A; or, if another, of type B. As far as I know, these statements are true, regardless of any variation in the reader's environment from fertilisation onward. The relevant genes have a direct, uninterrupted influence on a clearly defined aspect of blood chemistry.

Early modern genetics depended on discoveries of this kind. Many years of meticulous experiments and critical analysis led to the familiar concept of a chromosome as a stretched out necklace of genes. (At the time, what the genes consisted of,

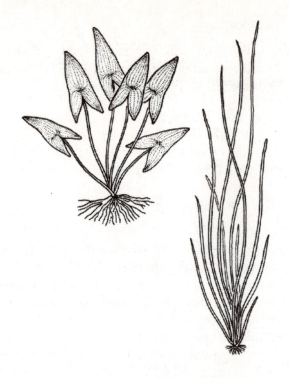

Not two quite different species, nor even a genetical difference. The arrow-head, *Sagittaria sagittifolia*, grows in ponds, where it has broad leaves; but in running water the leaves are narrow—an example of variation in response to the demands of different environments. The *phenotypic* difference is in this case environmentally determined. (After D.J. Futuyma, *Evolutionary Biology*)

and what they were threaded on, was obscure.) Often, the experiments required close scrutiny of small vinegar flies (*Drosophila melanogaster*). Features such as eye colour and bristle length were minutely observed. Similar experiments were done with domestic mice, of which we have many distinct types, and a garden plant, the snapdragon (*Antirrhinum*), which too is conveniently variable.

Evidently, the regularities expressed in 'Mendel's laws' hold throughout the animal and plant kingdoms. The existence of genes was inferred from the results of breeding. Each gene was presented as having an independent action on development. But, as we now see, such simple relationships are exceptional.

ACTUAL GENOMES

Much of the early work was summarised in 1919 by an out-standing American geneticist, T.H. Morgan (1866–1945). His book, *The Physical Basis of Heredity*, begins:

> That the fundamental aspects of heredity should have turned out to be so extraordinarily simple supports us in the hope that nature may, after all, be entirely approachable. Her much-advertised inscrutability has once more been found to be an illusion . . .

Perhaps Morgan, had he lived for another eight years, would have felt that sentence further justified; for, in 1953, Francis Crick and James Watson published their transforming first announcement of the discovery of the structure of DNA—the material of heredity in the cell. This led to a seemingly simple account of the chemical nature of genes and their relationship with cell proteins.

The DNA molecule, however, is not simple. On page 38, I remark that all the molecules of a given chemical substance are identical. DNA is different. The nucleus of a human cell (five to eight thousandths of a millimetre in diameter) contains about two metres of DNA. Each of us ('identical' twins perhaps excepted) has an individual pattern of the chemical units which make up these enormous molecules. Hence the reader's DNA is slightly unlike that of all others.

Nor is the action of DNA simple. Since 1953, study of the operations of DNA have revealed complications at an increasing rate. Some of the tricky features of modern genetics are sketched below. They justify the statement that the accounts of genetics which reach the public are often not only simplified but actively misleading. A reader who is willing to accept this without going into detail may skip the next passage.

✳✳✳

DYNAMIC GENES AND GENOMES

- In a complex organism the development of each fully formed trait, from egg to adult, requires a long series of steps. The total pattern of traits makes the phenotype. The phenotype of a human being is the whole individual. It must be distinguished from the genotype—the individual's collection of genes. Differences in the phenotypes of individuals may reflect genetical differences between them or they may be due to differences in the environments in which they developed. Commonly, they are due to both.

- The primary action of genes is to code for proteins. During development some, called housekeeping proteins, appear in all cells: they are encoded by housekeeping genes. But the human body has about 250 kinds of cell, each with its distinctive chemistry. The genes which code for special proteins, such as the haemoglobin present in red blood cells, come into action only in particular kinds of cell: in other cells, they are switched off.

- Gene action can be switched on or off by external (environmental) events. One of the functions of liver cells is to release glucose (a source of energy) into the blood. During vigorous exercise or starvation, hormones (glucocorticoids) from the adrenal glands stimulate genes in the liver cells to produce more of certain proteins; as a result, more glucose is secreted by the liver cells.

- Genes, as described above, are often thought of as responsible for single characteristics, such as a person's blood group or a fly's eye colour. But a single gene usually has multiple effects on development (pleiotropy). Hence, when a gene mutates, the result is usually change in many features. Some of the clearest examples are from ill effects of mutation. In achondroplasia, a mutant gene prevents the normal development of cartilage throughout the body; and, in addition, the circulation of blood in the lungs is defective and the teeth are distorted.

- The action of a gene in development can be enhanced or diminished by other genes (epistasis).

- Some interactions are on a large scale. Occasionally, a child is born with Down's syndrome (trisomy 21): growth is retarded; the face has an unusual appearance and the form of the hands is distinctive. The child has an extra chromosome in each cell. This chromosome is itself normal but it produces abnormal development. The conventional account of Down's syndrome includes a reference to impaired intelligence, but such people can live normal lives: their intellectual development depends greatly on upbringing and so illustrates, once again, the interaction of nature and nurture.

So genes do not act independently: all the genes in a nucleus (that is, the whole genome) combine to function as a unit. The idea of the gene as *the* unit of heredity is highly misleading.

- The presence of an extra chromosome, or the absence of a chromosome which ought to be present, is an example of large scale mutation. Very small, local mutations are much more frequent. Usually, they are harmless: the damage is repaired, inside the cell nucleus, by processes analogous to wound healing. All cells have a large number of proteins concerned with repair. But, occasionally, repair itself goes wrong. Xeroderma pigmentosum is a rare disease (a recessive condition) in which the presence of two copies of a mutant gene prevents one kind of repair. The result is a severe abnormality of the skin. On exposure to the sun, cancer may follow, because the light itself induces additional mutations in the exposed cells. (The development of a cancer usually requires a series of changes. Often, a change occurs only in response to something in the environment.)

- Only part of each chromosome consists of genes, that is, of DNA which codes for cell proteins. When, in 1977, this was first discovered, the 'noncoding' sequences were brushed off as 'parasitic' or 'junk' DNA. No longer. Much of it consists of gene-like units called jumping genes (transposons). These break a rule which once dominated genetics, namely, that all the active units in a chromosome are in a fixed order. Jumping genes are prominent in the genomes of Primates, including ourselves. When they move, they can cause rearrangement or mutation of ordinary genes.

- Like viruses, which they closely resemble, transposons can move between cells and even between different species. They are a recently discovered and highly disconcerting cause of genetical variation.

- Another rule has been called the central dogma of DNA genetics. It is expressed like this: DNA → RNA → protein. In the cell, genes, that is DNA, give rise to RNA; and the RNA in turn promotes the synthesis of proteins. The causal sequence is one way. This rule too was soon found to be misleading. Some of the offenders are lengths of DNA called retrotransposons. A retrotransposon moves

> by being first transcribed into RNA. A DNA copy of this
> RNA is then made; and this can be inserted back into
> another part of the genome where it can influence the
> action of ordinary genes. Retrotransposons are exception-
> ally prominent in the human genome.
>
> (For more on human genetics see Philip Kitcher's *The Lives
> to Come*, which deals also with the moral dilemmas resulting
> from modern findings.)

STABILITY AND INSTABILITY IN DEVELOPMENT

The achievements of modern genetics have reinforced fascination with 'the gene', but they have also worsened the chronic
confusion about what is 'inherited'. For gene action (outlined
above) is not simple. To sum up the complexities: change in a
single gene is almost certain to have multiple (pleiotropic)
effects on development; the genes also interact among themselves; hence the whole set of an organism's genes (the genome)
makes a unit. Moreover, in a chromosome, the genes are inter-
spersed among noncoding lengths. These include jumping
elements (transposons) which move within and between the
chromosome and even between cells; sometimes they move out
to other organisms. The movements of transposons can make
genes mutate.

Above all, *this genetical activity goes on in a changing environment.* The effects of the genes in a fertilized egg interact with
the surrounding conditions. For a human being, these include
the mother's uterus and milk, the cradle, schooling, disease
organisms and much else. This is the interaction of nature and
nurture which gives so much trouble.

Unfortunately, it is easy to think of a trait as due simply to
a person's genes, and not to a complex interaction. Consider
that widespread source of anxiety, obesity. Fatness 'runs in
families' and so seems to be 'inherited'. Sometimes, headlines
appear in the press: GENE FOR OBESITY FOUND. An
article follows, on the theme that fatness is 'genetic'. But a closer
look reveals that the 'gene' has been discovered in mice! Its
relevance to human fatness is, at best, tenuous. (A similar 'gene
for' fatness is now reported in human beings.)

Nobody doubts that people, like mice, differ genetically in the ability to grow fat. Fatness, however, is not a property of one's DNA: it is, again, an outcome of an interaction between genes and environments during development. For most practical purposes, the important knowledge concerns environmental factors which influence obesity. These include diet and other features of the life style (further described in my *Science of Life*).

From fertilization of the egg onward, development is strictly regulated and so produces a stable adult organism. But not all features develop predictably. As the case of obesity shows, some are inconstant or unstable in development. Muscles, like adipose tissue, grow or decline with use; immunity to disease is acquired or lost; and so on.

Instability in development is most obvious in our behaviour. The brain and the rest of the nervous system not only grow with elegant and improbable precision: they also enable us to vary our conduct, that is, to adapt what we do to changing circumstances. Survival often depends on such variation, that is, on intelligent action.

We not only display variation: we also provoke it. One of the achievements of modern behavioural science is the analysis of exploratory behaviour and of the demand for novelty. Healthy human beings are restless. Children especially, while awake, need incessant stimulation and, without it, complain of boredom. Much human action, once basic needs are satisfied, is directed to avoiding tedium. Movement, exploring, discovery of novelty and play are not only sources of enjoyment: especially in early life they also allow us to learn new things and—still more important—to develop our intelligence. Moreover, to some extent we choose our environments.

CLONES, HUMAN AND OTHER

Clones tell the same story. A clone of organisms is a group all of whose members have the same origin, hence the same genes. Cloning is an ancient horticultural practice: several plants, each grown from a cutting from the same parent, make a clone. ('Clone' is from a Greek word for twig.)

Another operation, also called cloning, is to take a newly

fertilized egg cell (ovum), to remove its nucleus by micro-manipulation and to replace it with the nucleus of a cell from another organism. Strictly, this elegant and highly skilled procedure should be called nuclear transfer. If the ovum so treated develops, the outcome is an organism with the same nuclear genes as the organism which provided the transferred nucleus. (Other, cytoplasmic genes are outside the nucleus and are not transferred. They, however, are few compared with the enormous number of nuclear genes.)

This is what happened to a notorious sheep named Dolly (now a mother), in an impressive technical development which may eventually have much use in stockbreeding. Perhaps some future flocks of sheep, herds of cattle and others will be clones of an ancestor with selected features. How far these features are sufficiently reproduced will have to be discovered by experiment.

Effective nuclear transfer is, however, difficult: 277 attempts were made before the success with Dolly. If the transferred nucleus is not at exactly the right stage, chromosomes may be lost or duplicated. Embryos then die early or produce monsters. During the experiments leading to Dolly, about 60 per cent of fetuses died in the uterus and others soon after birth. Ian Wilmut, the head of the Scottish research group responsible for Dolly, was nonetheless asked about the possibility of 'cloning' human beings. He is reported as saying that it might be possible; but he could think of no good reason for doing it. He is obviously a very sensible man.

True human clones are, however, an everyday phenomenon. The members of an 'identical' (uniovular or monozygotic) twin pair are derived from a single fertilized egg which, soon after fertilization, separates into two embryos. They therefore make a clone, of which each member has the same nuclear and cytoplasmic genes. Often, they are alike in appearance and hence have been popular in fiction. In drama, as in Shakespeare's *Comedy of Errors*, they have provided enjoyable farce. At the beginning of the play, we are told how the wife of a merchant became

A joyful mother of two godly sons:
And, which was strange, the one so like the other,
As could not be distinguished but by name.

In some other fiction, however, twins, though identical in appearance, differ in character: one is malignant, the other, virtuous. Tragedy may then follow. Is such diversity possible in real life?

Multiple human births, above two, are often clones. The Dionne quins, born in Canada in 1934, were shown by detailed examination to be derived from a single ovum. The first four cells of the ovum had separated and each had begun to develop on its own. Then one of the four separated yet again. As a result, the five girls each had the same set of genes and so made a true clone.

They were brought up together and treated and dressed in the same way. (They were also the victims of nauseating exploitation, for they were put on display for the entertainment of tourists.) Their early environment was as uniform as practicable. Their upbringing therefore left them little opportunity to develop independently. Yet, by the age of ten, as Amram Scheinfeld has described, each had a personality distinct from those of the others; and, as adults, they were all unhappy but in different ways. Two trained as nurses; two studied at university. Three contemplated a religious vocation but only one made it a career. One, an epileptic, died at the age of twenty; another died aged thirty-six. The rest survived into old age.

This outcome, by itself, ought to silence talk about recreating particular persons by nuclear transfer. The differences between the Dionnes are surprising only if one assumes that a fertilized ovum, or its DNA, already has a distinct personality. But of course it has not: personality depends on individual development. We each grow into a human being, perhaps rather similar to our parents; but we each also become individuals different from all others. From before birth, the reader has had his or her own experiences: even if they are quite like the experiences of relatives, they are unique to the reader. At the age of five or fifty, even if you have a twin, you are a person with memories and predispositions which belong to you alone. The distinctive features of your character are not in your DNA, though they are influenced by it. They are a product of your development in a varying environment; and they are partly an outcome of choices you have made. They cannot be reproduced by cloning. Genes are not destiny.

CHAPTER 6

HUMAN DESTINY

I returned, and saw under the sun, that the race
is not to the swift, nor the battle to the strong,
neither yet bread to the wise, nor yet riches to
men of understanding, nor yet favour to men of
skill; but time and chance happeneth to them all.

Ecclesiastes

W<small>E CAN NOW RETURN</small> to the question asked in chapter 3: if the human species is a product of natural selection, to what extent can biology explain what we get up to? Are we helpless flotsam on the sea of 'time and chance' that distressed the prophet? Or are we a species—the only one—with societies (or cultures) which we create or influence for ourselves, by intention and—partly at least—using our intelligence? Much of the answer has to come from outside biology, for we have to face exaggerations of the scope of science and neglect of what sets the human species apart from the rest of nature; we meet again the use of biological ideas to provide substitutes for religion and to support political programs; and we have to acknowledge the importance of history.

NATURAL SELECTION AS THE 'ABSOLUTE'

In *The Advancement of Learning*, Francis Bacon put forward the true ends of scientific activity as the glory of the Creator and the relief of man's estate. That was in 1605. In Europe, during the centuries that followed, doubts concerning an all-pervading deity grew; and, after 1859, scepticism was further encouraged by Darwinism. Many who tried to grasp the origin, purposes and future of humanity felt themselves to be left stranded, without a unifying principle or an overriding source of wisdom and morality.

Some philosophers therefore resorted to an idea called the Absolute. Before Darwinism, a strangely influential German philosopher, G.W.F. Hegel (1770–1831), had already proposed the notion of an Absolute Idea which was pure thought thinking about pure thought. This, according to Bertrand Russell, 'is all that God does throughout the ages'; which makes it, he adds, truly a professor's god. But later, another professor, an English philosopher, F.H. Bradley (1846–1924), did not quite agree: for him, the Absolute is not the god of religion because it has no personality. It is pure consistent thought which (I think) embraces everything, including god, and on which all else depends. Fortunately, I need not pursue this train of thought, even if I could, for it did not catch on for long. It does, however, illustrate the longing for an ultimate foundation on which all our ideas and practices can be built.

In place of the Absolute, some turned to biology and made from natural selection a transcendental principle which is still influential. While the Absolute was being debated, Herbert Spencer was presenting his concept of evolution as a cosmic process (page 42). Spencer was an early 'ultradarwinian'. For him, natural selection not only drives organic evolution: it is also the prime source of morality and of human progress. His slogan, 'the survival of the fittest', entered everyday speech. Perhaps contrary to his intentions, it was joined to the concept of *Homo pugnax* to provide the alarming portraits of a belligerent humanity described in chapter 3.

Later, others followed, but without the violence. Among them was a notable zoologist and readable populariser, J.S. Huxley (1887–1975), grandson of T.H. Huxley. Julian Huxley, a man of great charm and energy, did research on relative growth and animal behaviour. Then he turned to evolution and found in it, like Spencer, a source of morality. In 1953 he wrote that it could give rise to a new religion, Evolutionary Humanism. The evolutionary process in nature was held out as telling us what to do in the future. At the same time, in total self-contradiction, 'man' could now 'see himself as the sole agency of evolutionary advance': we (humanity) were to apply our own morality to Nature and guide the future evolution of organisms on the planet.

But Huxley's only practical proposal was for eugenic improvement. This was in the tradition of earlier political Darwinism. If Natural Selection is the Ultimate Good, it is wrong to interfere with it by measures such as national health services and even public education: these encourage the breeding of the poor, unfit and criminal and lead to social degeneration. The genetically best people should be the ones who breed. Obvious difficulties are blandly disregarded: how are the genetically best to be identified—best for what? who is to decide? Should breeding couples be forcibly matched, like cattle or camels, in genetical qualities? Above all, regardless of genotype, some environments are greatly superior to others, both for health and for developing ability. If children are being reared in an unfavourable environment, the environment can be changed. This objective has the merit of being quickly attainable; and, during the present century, in some countries it has been partly attained.

Our next step is to see what current evolutionary theory truly has to offer.

THE HORN OF THE RHINO: AUTHENTIC BIOLOGICAL SCIENCE

In 1966, an American evolutionist, G.C. Williams, produced a work, *Adaptation and Natural Selection*, designed to promote the progress of evolutionary theory (which it did). Williams begins with a statement about the scope of natural selection. His book, he says, 'is based on the assumption that the laws of physical science plus natural selection can furnish a complete explanation for any biological phenomenon'. Can this grandiose proposal be justified?

On page 75 I quote the geneticist, T.H. Morgan, as writing, optimistically, that modern genetics has simplified knowledge of heredity. Unfortunately, the concept of natural selection too, like genetics, can be introduced by simple histories. A disease organism, say, a bacterium, which at first quickly succumbs to penicillin, becomes immune and therefore more dangerous. Similarly, populations of mosquitos survive pesticides which were once effective. As well, the malarial parasites carried by the mosquitos become resistant to widely used drugs. Such adaptive changes, which are now many and alarming, are evidently genetically determined. They seem to be examples, on a minute scale, of evolution by natural selection.

But the notion of modern Darwinism as a complete theory, easily described, is another illusion. Simple presumptions about natural selection are always misleading. Its most conspicuous complication was familiar to Darwin. In *On the Origin of Species*, his most important work, he wrote on 'Correlation of Growth':

> I mean by this expression that the whole organization is so tied together during its growth and development that when slight variations in any one part occur, and are accumulated by natural selection, other parts become modified. This is a very important subject, imperfectly understood.

And here is J.B.S. Haldane (1892–1964), another giant of theoretical biology, in 1932.

> . . . with an animal or plant we are at first struck by its obvious adaptations; its claws, teeth, spines, protective colouring, and so on . . . But there remain a host of characters with no obvious value to their possessor . . . When we have pushed our analysis as far as possible, innumerable characters show no sign of possessing selective value.

As Darwin had realised, when a new trait emerges by natural selection, it can drag along with it other changed features which are not themselves advantageous. They have been called side effects, free riders or spinoffs. Yet it is still sometimes supposed that any feature, of any organism, is adaptive: that is, that it has or recently had selective value.

We can rarely say how the variation we see today has arisen. Darwin asked why the nest of a thrush (*Turdus philomelos*) is lined with mud, while the closely related blackbird (*Turdus merula*) prefers fibrous roots; he could find no answer. R.C. Lewontin in 1979 similarly pointed to the difference between the Indian rhinoceros and the African: why has the African species two horns, the other only one? Perhaps no 'Darwinian' answer exists: we need not assume that conditions in Africa once selected for two horns.

These are typical examples of what we find throughout nature. Evolutionists have therefore proposed two further limitations of natural selection: 'drift' and 'neutrality'. Small groups (demes) of a single species sometimes vary genetically, not as a result of selection but by chance or drift. Until a few thousand years ago, the human (or prehuman) population of the world seems to have been less than ten million—the size of some modern cities. Such a sparse population is just right for drift. Isolated groups can survive despite the presence of individuals with, at first, nonadvantageous genomes; they can later evolve (by natural selection) on lines separate from those of others. (Given much time, they can even separate into new species.)

A special case of drift is the founder effect. A single individual with an apparently disadvantageous genotype can originate a flourishing lineage. Porphyria is a rare disease, usually dominant, in which exposure to light causes severe blistering and a painful abdominal condition. Yet it has a high incidence among people of Afrikaner descent in South Africa. All these porphyriacs are descendants of a man who settled in Cape Town in 1686. Other such founder effects are known. No doubt, many more have remained unidentified, because they were not marked by a distinct condition.

The second theory, that of 'neutrality', arises partly from the high and uniform rate at which mutations occur. Some remain unrepaired (compare page 77). Perhaps many mutations have *no* initial effect on fitness. If so, genetical change due to drift becomes still more likely.

The theories of drift and neutrality do not reject natural selection: they propose modifications of the ways in which evolutionary change can occur. Both have led to controversy. Douglas Futuyma, in a helpful textbook, comments on the prodigious rate at which papers on neutrality are published, without achieving a consensus. Fortunately, neither the author nor the reader is obliged to take sides. We should merely note that evolutionary knowledge and theories are continually developing and that much remains to be learned.

For more on the complexities of evolution, consult S.J. Gould's essays, especially *Ever Since Darwin*, and for a technical account *The Shape of Life* by R.A. Raff.

How To Achieve Fitness

For genetical change to increase the chances of survival, the necessary genetical variation must first exist. This is rarely a matter of a single gene mutation. As we know, genes interact with each other and, in development, the effects of one gene can combine or conflict with those of others. Clashes must be avoided. It would be no good if a mammal in a cold climate developed such massive insulation, from fat or hair, that it could not support the extra weight. Growth of each part must be matched with that of all the others.

A famous case of relative growth (allometry) is the Irish elk (*Megalocerus giganteus*), the largest of all deer. Increase in total body size (presumably a result of selection) went with the development of disproportionately large antlers. Its extinction has been supposed to be due to the handicap of these huge structures; but, as S.J. Gould has shown, this is at best surmise. It is lamentably easy to guess, with little or no evidence, what has 'driven' evolution and what has led to the extinction of species.

Some exceedingly intricate relationships are with other species. The giant cowbirds (*Scaphidura*) of Panama are nest parasites, like the European cuckoo (*Cuculus canorus*). They lay their eggs in the nests of large orioles called oropendolas (*Gymnostinops*). The oropendolas often remove the cowbird eggs from their nests; but some cowbirds lay eggs which mimic their host's eggs and so are not thrown out. Some oropendolas, however, tolerate the presence even of nonmimetic cowbird eggs. This is explained by the behaviour of the cowbird nestlings: they mature early and eat the larvae of botflies which kill many oropendola nestlings.

Such weird interactions are certainly widespread. They give a glimpse of what we discover when we try to find our way in the labyrinth of evolutionary change. They make it hardly surprising that evolutionary transformations are usually very slow.

Nostalgia

When all the complexities are allowed for, the theory of evolution becomes obviously unable to explain either the origin of humanity or our present peculiarities. We cannot decide what features were crucial, in any period, for the survival of our ancestors. We cannot even be certain how our ancestors behaved, say, five million years ago. Perhaps preoccupation with

Fragments of our possible past

Species	million years ago	endocranial capacity, cm^3
Australopithecus		
ramidus	4.5	400?
afarensis	3.5	413
africanus	3.0	440
boisei	2.5 to 1.5	463
Homo		
habilis	2.0 to 1.8	640
erectus	1.5 to 0.5	930 (+?)
sapiens (modern)	0.1 to present	1450

The anthropologist, P.V. Tobias, has described paleoneurology as 'a field beset with relatively few facts but many theories'. The figures above, based on distressingly few fossils, reveal our probable ancestors, up to about two million years ago, as having brains similar in size to those of apes. Then came a large and rapid increase in cranial capacity. This coincided with the extensive manufacture of stone tools.

such questions reflects a desire for security: the past cannot be altered, therefore it is safe. In the words of a historian, G.R. Elton, 'The future is dark, the present burdensome; only the past, dead and finished, bears contemplation'.

As for our remote past, what do we know? Six million years ago, our possible ancestors, like chimpanzees and gorillas today, lived in African forests and seem to have been well adapted to moving in trees. A change of climate and a thinning of the woodlands may have favoured the ability to walk and run in the open on two legs. The hands then became increasingly free for elaborate tinkering. Eventually came the evolution of a grasping thumb and extreme dexterity. During the same period, a reduction in the size of the gut, the jaws and the teeth seems to have gone with a change from a diet largely of fruit and leaves to one which included grass seeds and animal food. And, after four million years, came a vast expansion of the brain.

Although the details of their life style are unknown, our ancestors were presumably exploratory and inquisitive (page 79). Our impulse (or instinct!) to search for novelty, like the habit of tinkering, we share with the living apes and, we may suppose, with our common ancestors of several million years ago. Stone tools (initially owed to an australopithecine genius?) altered slowly at first; but, later, our ancestors' manual skills led to increasingly rapid nongenetical changes.

Features of 'Human Nature'

Erect posture; arched foot; grasping thumb; free hands
Omnivory; small gut and teeth
Enormous brain
Loss of 'instinctive knowledge'
Extreme adaptability of behaviour
Tool making and fire
Speech and symbolism
Teaching and tradition
Morality

The transition from ape to human being took place during the Upper Paleolithic—a period which ended about 100 000 years ago. By that time our predecessors had learned how to control fire and so had begun a complete transformation of the human condition. Controlled fire and multifarious manufactured equipment have no counterparts among animals. Though universal, they are not instinctive but are transmitted by example and teaching. In the words of the Australian archaeologist, V. Gordon Childe (1892–1957), they are not part of 'man's body'.

> He can leave them and lay them aside at will. They are not inherited in the biological sense, but the skill needed for their production and use is part of our social heritage, the result of tradition accumulated over many generations, and transmitted . . . through speech and writing.

So *Homo sapiens* emerged with the features in the list above. From that beginning came the species with a history.

THE MYSTERY OF MEANING

Of the items in the list, a central feature is speech or language. *Homo sapiens* is also *Homo loquens*, the talkative species. Our ancestors of several million years ago must have communicated by grunts and cries similar to those of the living apes. Their replacement by words required changes in several parts of the brain. With them came a rapid increase in brain size. Both producing speech and also the ability to understand it demand neural systems of unravelled complexity (chapter 2).

As well, alteration of the mouth and larynx allowed consonantal sounds to be made. An ape may be able to exclaim, 'aah'.

A human being can say, 'part' or 'parrrt', both of which include the 'aah' sound. The ability to produce additional sounds (such as those represented by 'p' and 't' and the rolled 'r') is acquired in early life: like other skills, it is learned; and the precise form of what is learned depends on the social environment (the culture).

A language consists of sentences. Animals do not utter sentences, or even words of the kinds used by human beings. Animal signals in nature indicate a particular situation: readiness to mate; discovery of food; the approach of a predator; and others. Prolonged and ingenious efforts to teach chimpanzees to acquire some equivalent of our speech have achieved little more than 'please give-me orange'.

Our spoken languages consist of only about thirty or forty distinct sounds; but, unlike animal signals, these are put together in tens of thousands of words. A sentence contains words or other symbols, arranged according to a system of rules. (In some languages, word order is crucial: in English, a Venetian blind is not at all the same as a blind Venetian.) These rules enable us to make promises and requests, to tell stories, to teach and to argue. A monkey may utter a call when a dangerous snake appears; but it cannot use the sound to refer to a snake in the grass—a metaphor for a treacherous or dangerous individual of the speaker's own species.

Some anthropologists believe that speech arose from selection for the ability to communicate thought. Others emphasize the conveying of social information and attitudes. Yet others suggest news (and teaching?) about tools or food. Perhaps all these had a selective influence at some stage. It is impossible to tell. Nor can we decide to what extent incidental effects, such as correlated variation or drift, played a part.

What we certainly know is what we observe here and now. Yet worldwide search has revealed no traces of primitive or early speech in any of the four thousand or so extant languages. All are fully developed (and make heavy demands on translators). The languages of some tribal people, such as Australian Aborigines or the North American Cherokee, are even more complex than (for example) English.

Since speech is a universal feature of the human species, it has been called instinctive. Steven Pinker has indeed written an

A Magdelenian masterpiece. Cave painting of a man wearing a stag mask, from about 8500 BC.

interesting book entitled *The Language Instinct*. But our incessant talking, like other social features of *Homo sapiens*, falls outside any conventional biological category. In the beginning, the ability to speak was no doubt at least partly due to natural selection; but, once evolved, it was followed by the development of immense diversity among groups and individuals. This variety reflects our ability to create traditions and to learn from example or instruction: it is not genetically determined.

Speculation about the evolutionary past (phylogeny) is fun but it can distract us from attention to the development of speech in early life (ontogeny), hence from findings which can be put to good use. Individual development, unlike our past, can be studied by direct observation and experiment. The sensitive period for learning speech is from just over a year to six years. At these ages, children are, in this respect, far ahead of their parents: they learn the rules which prevail in their own community; and, even more remarkable, each person develops his or her own voice and style of speech (and later, if literate, of writing). The communication skills of most of us remain, however, notoriously imperfect: to improve them we have to exert ourselves throughout life. The fruitful questions therefore concern, not how the human species acquired speech in the remote past, but how we and our children learn to communicate with others.

SOCIAL EVOLUTION

When human beings appeared, presumably already talking, about 100 000 years ago, their skulls resembled ours; so, no doubt, did the neural structures concerned with language. Since that time, and especially recently, prominent changes in human action (but not bodily structure) have occurred at increasing rates.

Unless they are in dire straits, human beings have much time left over from activities essential for survival and breeding. For at least 40 000 years, gatherer hunters have had spare time not only to make elegantly crafted tools but also to create drawings and paintings on stone and bark. The latter were not maps of how to find water or food (nor were they for sale): they seem to have been works of art or of magic. The exertions needed for producing them in caves, by the dim light of flickering oil lamps, must have been prodigious. The techniques required, such as mixing pigments, were no doubt passed on by teaching.

A little more than ten thousand years ago, food production replaced gathering and hunting, apparently independently, in several regions. Each region offered animal species which could be domesticated. Agriculture began in the fertile crescent of the 'Near East', which includes parts of Turkey, Syria and Jordan. There, at about 8500BC, the important animals were sheep and goats. The plants were wheat and barley (grasses); peas and lentils (pulses); and olive. Later, in China, pigs, rice, millet and soybean fed an increasing human population. In Central America people domesticated corn (maize) and the turkey; and in Amazonia, potato and the llama.

Later still, a settled life allowed people to construct gigantic stone monuments. The most famous, Stonehenge, in the south of England, is 5000 years old and is only one of hundreds of similar age. Among the others are the imposing works of the

Stonehenge as it may have appeared when complete. This, the most famous of many prehistoric stone monuments, combined a knowledge of astronomy with religious practices. (See David Souden's *Stonehenge Revealed*.)

Maya, in Central America. The earliest known, discovered in southern Egypt in the 1990s, dates back 7000 years.

The architects were knowledgeable astronomers: without written traditions, they fixed the sun's position at sunrise or sunset at the solstices by alignment between two pillars of different heights. But their work went far beyond what was needed for a calendar. The outer circle of Stonehenge consists of enormous stones, harder than granite, which had to be moved—by the exertions of hundreds of people—from the Welsh mountains, pounded level by hand and then raised by astonishing feats of engineering. A modern mason with a stone hammer took an hour to remove six cubic inches. Three million cubic inches (or more) were removed, which represents at least half a million man hours or 250 man years.

A grave near Stonehenge contains a skeleton, a stone mace, axe, spear and daggers. A historian, Tom Shippey, suggests that this was one of the men who directed the masons and who derived their authority from knowledge of the heavenly bodies.

All through history and prehistory, small communities, poor by our standards, have similarly lavished resources on works of art and religion. Stonehenges and Greek temples, like paleolithic stone implements and neolithic cave paintings, were produced by people who must have transmitted their knowledge and their appreciation of the fine arts by example and teaching.

Two recent authors have surveyed the long history of human knowledge and skills during and after the rise of agriculture. Both emphasize, as universal traits of humanity, curiosity and the capacity for invention. Jared Diamond, in *Guns, Germs and Steel* (a misleading title), writes especially on the production of food. George Basalla, in *The Evolution of Technology*, is concerned principally with hardware.

Diamond not only describes social evolution but also tries to explain it. His explanation is, at least in part, ecological. A leading theme is the origin of agriculture and the strange differences between human societies in different regions. Why did the Aborigines of Australia, New Guinea and southern Africa not become as technically advanced as those of Europe and Asia? During the nineteenth and early twentieth centuries, Europeans often assumed their natural superiority over these 'primitive' people (chapter 3). Superiority was equated with technological development, expressed by the cynic who said, 'We had the gatling gun, and they had not'. But, as usual, no good ground exists for supposing such groups to be genetically inferior. Inferior in what respect? Diamond, indeed, suggests that survival without agriculture, or only with simple farming and fishing, requires greater ability than living a sheltered life in a technically advanced community.

The world's distribution of animal and plant species is very uneven. The whole of the vast area of Australia offered its aboriginal inhabitants nothing equivalent to wheat, maize or the potato, no counterparts of cattle, sheep, horses or camels. But, once introduced, all these have thrived and Aborigines have adapted to them. Hence it is possible to interpret some human diversity from knowledge of what species could be domesticated. In this, we are helped neither by evolutionary theory nor by genetics.

The same applies to the technological advances described by Basalla. His account begins with a reference to Darwin and natural selection; but, as he shows, organic evolution is completely different from technical change. Here are some familiar instances. When, in the eighth century, printing reached Europe from China, that was not a result of natural selection; nor was the arrival of gunpowder in the tenth century. In the opposite

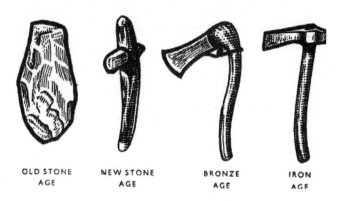

| OLD STONE AGE | NEW STONE AGE | BRONZE AGE | IRON AGE |

The 'evolution' of the axe. An example of socially determined changes in hardware. Such modifications occur at increasing rates in our own period. The hectic time scale is totally unlike that of organic evolution.

direction, the British, mainly during the nineteenth century, presented India with railways, steamboats and the electric telegraph. These were not automatically selected random events: they were planned.

Often, it is the details that interest us. In the sixteenth century, Portuguese adventurers found their way to Japan. Two of their muzzle-loading guns (matchlocks) were bought by Japanese and were copied by Japanese metal workers. The copies worked regrettably well and for a time had an influence on local wars.

Sometimes, copying is resisted. In the sixteenth and seventeenth centuries, religious persecution drove many Protestant Huguenots from France. With them went valuable skills in the manufacture of textiles. Early in the eighteenth century, English textile merchants, anxious to adopt a technique of weaving silk kept secret in Italy, sent a spy; and he, after two strenuous years, learned the method and brought it back to London. In this famous case, we have an indirect influence of religion on the spread of technology as well as an early instance of industrial espionage.

These examples of the 'evolution' of technology, and its social reverberations, may seem incongruous in a short book on science. That is why I include them. Actual case histories are needed to bring home the message, that we cannot interpret our social lives solely by resort to biology.

THE RENEWAL OF ART AND SCIENCE

For at least forty thousand years, the growth of technology has gone with the practice of the fine arts. A succession exists, hardly broken, 'from cave painting to comic strip'. (That phrase I owe to Lancelot Hogben.) All children can learn to draw, paint and model. This faculty (like speech) evidently originated during the emergence of the human species. But, since then, the arts have diversified to a bewildering extent and have become sources of both joy and controversy. And artistic skills and appreciation, like those of music (pages 67–8), have to be taught.

So this history too can be studied only as something distinctively human. In almost all periods artists have pictured people or animals. In his celebrated *Art and Illusion*, the historian of art, E.H. Gombrich, asks why, at different times and in different places, the creations of artists are nevertheless so distinctive. His question resembles Diamond's problem above: why did only some people develop agriculture and an advanced technology? Although, like Diamond, Gombrich can give no final answer, he knows that biological analogies cannot help him. As he says, in histories of art 'evolutionism is dead'.

One of Gombrich's themes is the stability displayed by the distinct styles of some countries and ages. The Pharaonic art of ancient Egypt and the Buddhist art of India each maintained traditional forms, especially those representing persons, for many centuries. So did the art of the European Middle Ages.

The art of both Egypt and the Middle Ages also literally lacks perspective: hence follows the top heavy appearance, in many paintings, of woods and gardens. This went with a lack of 'perspective' among historians. In the Middle Ages, neither historians nor artists worried much about anachronism. The people in a painting of an event dating from ancient Rome often wore medieval dress.

In Europe of the late thirteenth and fourteenth centuries, that attitude was discarded. The West began an upheaval in both the arts and the sciences, the Renaissance, which is still going on. In art, a leader was the Florentine genius, Giotto di Bondone (1266–1337). One of his daring innovations was to paint people as distinct persons, suitably dressed and expressing emotions appropriate to the events he portrayed. So Giotto

Lack of perspective: an ancient Egyptian drawing in which the top heavy appearance is due to the artist's inability to represent the effect of increasing distance. Contrast the drawing on page 123. (From R.L. Gregory, *The Intelligent Eye*)

helped to create a demand for an art which clearly represents both actual people and recognizable stories; or perhaps he was responding to this demand.

In the fourteenth and fifteenth centuries, a group of Florentine artists went further. Among them was Filippo Brunelleschi (1377–1446). Accurate representation demanded not only recognizable portraits but also geometrical correctness. Brunelleschi was the principal founder of modern architecture, and seems also to have originated the modern understanding of perspective. (Some authors write of 'scientific' perspective but in fact it was a mathematical technique.)

This liking for formal accuracy and geometrical precision makes a paradox, for a constant feature of the Renaissance was an emphasis on innovation. Since then, the history of the arts in the West has been one of incessant experimenting: tradition has become not a source of rules about what should be done but, as Gombrich writes, 'the starting point for corrections, adjustments, adaptations'.

At this time another kind of exact representation, combined with experiment, had a beginning: modern science. One of its founders was a Rhinelander, the son of a fisherman, who became a cardinal. Nicolas of Cusa (Nicolaus Cusanus, 1401–1464) was a mathematician who is credited with the first biological experiments of the modern period. His quantitative observations on growing plants depended on the innovative use of the balance for experiments, on which he wrote a book. He also rejected the earth as the centre of the universe: the cosmos, he said, had no centre. Still more remarkable, he anticipated the philosophers of our own time who hold that it is possible to approach closer to the truth but never to achieve finality. (He even opposed the persecution of witches by the church.)

Although an original thinker, Nicolas himself avoided persecution, evidently because he wrote at great length on theological questions. This may also have reduced his impact on the development of early science. He did, however, influence Giordano Bruno (1548–1600). Bruno entered a monastery as a young man but quarrelled with his superiors, was suspected of heresy and had to take flight.

Before Bruno's time, Copernicus (1473–1543) had proposed an astronomy in which the sun was the immovable centre of the cosmos, around which all else revolved. (By deferring publication to 1543, the year of his death, Copernicus avoided the attentions of the Inquisition.) Bruno combined the proposals of Nicolas and Copernicus and so was led to the familiar modern concept of the earth moving round a sun which itself moves. But Bruno had, according to Charles Singer, a 'restless and turbulent spirit' and a 'lofty indifference to the dictates of common sense'. As a result, in 1593 he was imprisoned by the Inquisition; and in 1600 he was killed by burning.

MUDDIED MEMES

Although the many facets of social evolution are complex, they are not obscure. Yet some prominent writers are so fascinated by Darwinism and genes that they neglect them. Earlier, I quote G.C. Williams suggesting that the laws of physical science plus natural selection can explain any biological phenomenon (page 84). But he also writes:

> I cannot readily accept the idea that advanced mental capabilities have ever been directly favored by selection. There is no reason to believe that a genius has ever been likely to leave more children than a man [or woman?] . . . below average intelligence.

And he shows foresight when he adds that 'a biologist can make any evolutionary speculation seem scientifically accept-

history. Certainly, understanding *human* nature requires historical knowledge. Without it, it is easy to fall into the trap of biological naturalism. A particularly clear instance of error, due to ignorance of history, is the equation of human landholding with animal territories (page 48).

When, however, we ask historians themselves what they have to say, we find them differing on the nature of their subject: like biology, historical studies are various. Gordon Childe opens his review of historical order by taking technology, or 'the tools and machines of production', as of primary interest. Like Diamond and Bacalla, cited earlier, he concentrates on a natural history of humanity.

D.J. de Solla Price, an American historian of science, rather like Diamond, writes of 'the desire to dig deeper and find out why and how things happened the way they did'. But his concern is with detail. He describes how his method leads to the breaking of idols: for instance, contrary to a popular story, 'Galileo almost certainly did not make an experiment [on falling bodies] from the tower of Pisa'. For another, more formidable breaker of idols, Karl Marx (1818–1883), the key to history is study of the conflicts between economic classes. This, he says, can help people to use science to remedy human ills. Similarly, for the English Catholic, Lord Acton (1834–1902), who has been called one of the great Victorian misfits, knowledge of the past 'is eminently practical, as an instrument of action and a power that goes to the making of the future'.

Impersonal histories, however, omit much. An English political historian, E.H. Carr (1892–1982), holds that 'history cannot be written unless the historian can achieve some kind of contact with the mind of those about whom he is writing'.

All these scholars show ways in which history increases our understanding of ourselves. Few attempt general statements such as the 'laws' of nature sought by scientists. Instead, they interpret historical events with the knowledge that people have needs and desires and (sometimes) 'godlike reason'. Their concern is usually with particular happenings. Occasionally, they comment on how their subject is related to the sciences. One, G.R. Elton, unexpectedly suggests that history is more objective and independent than science, because its subject matter is fixed: historians cannot alter their material by experimenting. Even if we disagree about objectivity, we must accept that science has a history which is not itself part of science.

Yet some modern writers, with a narrow focus on biology, evidently regard historical studies as pointless or trivial—an attitude, sometimes called scientism, which treats scientific knowledge as superior to all others. A series of broadcast lectures by an eminent Australian historian, Manning Clark (1915–1991), seemed to acknowledge this outlook. He began with a putdown by a famous novelist: he was aware, said Clark, of the stern warning by L.N. Tolstoi (1828–1910), that historians are like deaf people who go on answering questions that nobody has asked them. Clark responds, however, that all human beings have to find a way in which the world becomes intelligible and bearable; and one means of doing so is knowing about our history.

WHAT DO WE KNOW?

I once asked Manning Clark what he thought about the different ways of writing history. He replied that he did not bother with them. Perhaps this was his way of saying what is certainly true: that we can learn from them all. Whichever we prefer, when it is suggested that biological theory can replace the labours of historians, both biology and history are obscured. To force all the complexities of human intelligence and stupidity, skill and ineptitude, friendship and enmity, pacifism and violence, altruism and egoism, love and lust, peace and war, into a 'Darwinian' or 'instinctivist' frame is to ensure that we fail to understand them. Worse, such an attitude implies that we cannot usefully try to improve the human condition. Yet the achievements of the sciences, in our time, without any other considerations, clash with such pessimism. The second half of this book gives more reasons for dismissing such melancholy.

I now end this, the first half, with a summary, both of what we know and also of what we do not know, concerning human evolution and the influence of nature and nurture on how we develop.

- The human species is presumed to be a product of evolution by natural selection; but the complex of features which were, at any period, crucial for the survival of our ancestors cannot be identified. Our swollen brains evidently reached their present state many tens of thousands of years ago. How they did so is a matter of speculation.

- We now exercise a great variety of skills, manipulative, verbal, mathematical and other. Not all can justly be supposed to have carried a selective advantage over alternatives. Much human activity, notably in the arts and in play, is made possible because we have spare time when we have done everything needed to ensure survival.

- Human beings are dexterous, articulate, curious, restless, speculative and argumentative. We are not, as are nearly all other species, adapted for a single, clearly defined way of living: instead, we are adaptable without visible limit. As a result, once the human species had evolved, social changes increasingly occurred at rates much greater than those of biological evolution. These changes, which are still going on, have included technical innovations, new developments in social organisation and in the fine arts and music and novel ideas in realms ranging from mathematics to morals.

- Such changes are not genetically determined, nor are they consequences of 'random', unplanned variation: they depend on teaching and imitation and are due in part to decisions made by thinking people. They represent intentions which are often made long in advance of the planned outcome. They are the material of history and belong in a dimension of knowledge outside that of conventional biology.

The form and personality of the reader and the author, as of other human beings, reflect three sources of variation.

- First, we—each one of us—differ genetically from everyone else (some twins perhaps excepted).
- Second, we—each one of us—develop in diverse surroundings which, at least in detail, differ from those of all others. During development from the fertilized egg, we are continually interacting with varying environments. Many of these external influences are cultural.

- Third, we—each one of us—by acts of will control, to some extent, the conditions in which we and our successors live. Hence follow our moral obligations to our fellows.

The first two items above, which often cause severe difficulties of understanding, belong to biology. The third comes not in the natural sciences but in metaphysics and ethics. It is, however, an axiom which, as we go about our daily lives, is inescapable. Some of its implications appear in the rest of this book.

PART III

'THE SCIENTIST':
THE IMAGE AND THE REAL

How are scientists imagined? And what are they really like? Some of the answers commonly given are obstacles to understanding science. But not all: a classical scholar, Maurice Bowra, once complained that scientists are treacherous allies on university committees, because they are apt to change their minds in response to arguments. Was this compliment justified?

CHAPTER 7

MAGICIAN, EXPLORER, TECHNICIAN OR BORE?

There are whose study is of smells,
  And to attentive schools rehearse
How something mixed with something else
  Makes something worse.

. . .

Others the heated wheel extol,
  And all its offspring, whose concern
Is how to make it farthest roll
  And fastest turn.

QUINTUS HORATIUS FLACCUS
(parodied by R. Kipling)

AT LUNCH ONE DAY, while I was writing this book, I asked
three people how they saw scientists. One, a mathematician,
said, 'As bores'; but he hastily added, 'Mathematicians are worse'.
When I turned to a musician, she said, 'As seekers after truth'.
But a physician said that she had found them very friendly:
'flirtatious, even'. This was hardly a well designed survey; but
it does give a glimpse of the variety of attitudes to scientists
and their work. Another is implied in the epigraph above.

IMAGES

In systematic surveys, when people are asked how they see
scientists, the answers are as mixed as the replies offered by my
friends. According to one account, the surveys yield 'an almost
wholly negative estimate'. Proverbial speech gives us plenty of
images of types of people: drunk as a lord, sober as a judge,
mad as a hatter and so on. To them we should perhaps add
'cold as a scientist', for some surveys present us as unapproach-
able, unsociable, irresponsible and narrow. (If scientists are
asked, they are found to regard themselves as the exact opposite
of all these.)

'The scientist': a child's portrait. (Courtesy Roslynn Haynes)

When young children have been questioned, they have portrayed scientists as men (no longer quite right); either as eggheads or with untidy hair (quite right); working alone in a laboratory (quite wrong); commonly on secret or dangerous projects (mostly wrong); and as a source of anxiety (quite right).

The phrase 'negative estimate', quoted above, is from a work published in 1994 by an Australian professor of English, Roslynn Haynes. Some later news is better. In 1996, surveys published in the USA and Britain each showed a high level of confidence at least in science, if not in scientists. And, in 1998, in a British poll, scientific (and other) professors scored highly on truth telling—far, far above politicians and journalists.

The book by Roslynn Haynes is mainly a collection of fictional portrayals of scientists. These are alarming in more than one sense: the imagined scientists themselves are often frightening; and the portraits are grim enough to worry any real scientist. At worst, invented scientists become nightmare figures. In a play, *The Physicists*, by Friedrich Dürrenmatt, the principal characters are grotesque caricatures named Newton, Einstein and Möbius. The action takes place in a lunatic asylum where the physicists are patients. In the first act they murder their nurses; in the second, they are revealed as secret agents. An eminent American physicist, Freeman Dyson, describes how he once complained about the unreality of the characters to a colleague. The friend replied that the whole point of the play is to show scientists as they appear to the rest of humanity.

A reader may prefer to dismiss that play as reflecting a writer's paranoia and overheated imagination. But, even if Dürrenmatt does go over the top, he represents a real social phenomenon. Eric Hobsbawm, a learned and humane historian, has surveyed the social history of the twentieth century. The century, he says, was not at ease with the science which was its most extraordinary achievement and on which it depended.

> The progress of the natural sciences took place against a background glow of suspicion and fear, occasionally flaring up into flames of hatred and rejection of reason . . . The idea that science equals potential catastrophe essentially belonged to the second half of the century.

As a result, we face portrayals of scientists as emotional cripples—inhuman researchers who think only in bare facts and numbers; we are also presented as arrogant, contemptuous of nonscientists and power-crazy. All this leads us into hideous blunders. In a 'prophetic' story, written more than a century ago, an inventor creates a chess-playing machine, then takes the machine on and wins. So the machine, in revenge, turns on the inventor and kills him. (I write this passage while the press is predicting disaster in the year 2000 or Y2K, owing to the collapse of the computers on which we depend. But these computers are not—I think—held to have human emotions.)

Worse, some women writers, justly upset by the fact that too high a proportion of scientists are men, attack scientists for cutting themselves off from the natural world: biologists, they say, treat Mother Nature as somebody who waits passively to be manhandled. Hence we appear as male chauvinists who treat Nature as a woman; and rape her.

So stories about science and scientists are dominated by misleading melodrama. They reflect the wishes or anxieties of the writers and of many others. These anxieties are not only, or even principally, biological. Spencer Weart has written an impressive essay on nuclear fear, in which he describes an 'image of scientists as children tinkering with forbidden secrets'. As Weart shows, portraits of inhuman scientists today arise largely from the employment of science for evil ends. These ends are desired by powerful elements in modern society but they are increasingly opposed. Some recent fiction does indeed reflect the need to protect the planet from untamed technology; but the writers rarely acknowledge the efforts of scientists who resist the pollution and destruction of the biosphere and hammer away at governments and industry to alter their policies.

Correspondingly, most fiction about scientists is written by people with no experience of science. The best fictional portrayal I know, of what actually goes on in a laboratory, is in *A Sort of Traitors* by Nigel Balchin. Near the beginning, two young research assistants are recording the results of experiments on the growth of a mould. In thousands of these cultures they nearly always find only 'slight growth'. They report their negative finding to the professor; and he decides that the whole procedure must be repeated with a slight change of conditions.

Of course, not all research is quite like that. And the slave-driving professor is himself carrying out valuable and successful experiments. But even the most exciting and fundamental discoveries require solid, painstaking, slow work. So it is not surprising that accounts of science and scientists in novels and plays are unreal: if authentic, they could hardly be thrilling or even readable.

THREE CARICATURES

Characters in fiction not only reflect common opinion: they also help to form it. Some have genuine relevance to what goes on in the real world. At least three significant caricatures of 'the scientist' exist: first is the magician; then, opposed to that, is the fussy, pedantic bore; last is the philistine technician.

The scientist as sorcerer descends from the ancient respect for the magi. These Persian priests (who gave us the word magic) could cure diseases by sorcery. (Compare the necromancers in chapter 1.) The most famous image of the scientist as magician is, however, embodied in Mary Shelley's notorious Victor Frankenstein. The achievements of her 'scientist' (a medical student!) were designed to 'awaken thrilling horror'. Frankenstein manufactured a living monster from the organs of fragmented corpses and so 'created life'. For the journalist Jon Turney, in his *Frankenstein's Footsteps*, this fantasy is the 'governing myth' of modern biology. Certainly, many of us have encountered, in movies, some incarnation of the monster. (But not all modern scientific wizards are enemies: for instance, the admirable 'Dr Who?' in television programs for children.)

If the wizard relies on an inner source of inspiration, he has links with the fictional science which consists of dramatic discoveries. According to a once popular idea, occasionally somebody rushes out on to the lawn, crying, 'Heureka!' (If one did that in fact, one's colleagues would send for a psychiatrist, not the Nobel committee.) A major omission in such stories is the activity which precedes discovery. The discovery itself may be exciting or only trivial; but before it is made comes the hard labour; and this, to repeat, commonly entails accumulating meticulously observed facts.

We are now approaching the second portrait: the bore. Collecting and classifying facts requires special skills and a high degree of dedication. But it is not in itself exciting. The nature and importance of such endeavours are often undervalued. An entomologist may devote a large part of his or her life to classifying, say, the more than five thousand species of Acrididae. Such work is often endowed by 'the taxpayer'. How scandalous, somebody will certainly say, that 'public money' should be squandered on studying obscure insects. But the Acrididae are grasshoppers; and among them are the locusts, of which a single swarm can totally destroy the farms and vegetation of a large area. (So can some grasshoppers which remain on foot.) The discovery of how locusts can be prevented from swarming—one of the most impressive achievements of twentieth century economic zoology—would have been impossible without the labours of unknown people, who pored over specimens in the small back rooms of universities and museums. Nobody could tell in advance which of the species they studied would be important.

The idea of science as boring has been promoted by some philosophers of science. Here is the strange opening sentence of a textbook on scientific method.

> The scientific mentality may be roughly characterized as the tendency to suspend belief until evidence of the appropriate kind is produced, and then to believe the proposition in question only if the available evidence warrants it.

This makes 'the scientist' sound like the notorious sceptic out for a walk with a friend. It is spring. The friend points to a flock of sheep crossing their path and remarks that they had been sheared. To which the sceptic replies, 'Well, on this side, at least'.

Granted, we do insist on sound evidence, or try to. But evidence about what? A reader of the quotation above might get the impression that all we do is hang about waiting for propositions or facts to turn up, and then begin nitpicking. No hint is given of curiosity or restrained guesswork, let alone inspiration. To repeat, while actually doing science, we try to find things out.

> 'Something hidden. Go and find it. Go and look behind the
> Ranges—
> 'Something lost behind the Ranges. Lost and waiting
> for you. Go!'

This, from Kipling's peom 'The Explorer', may seem itself rather overheated; but it is relevant, because the practice of science is an exploration—even if the exploring does require much pedestrian finding of facts.

Yet another, more important, source of the idea, that we are terrible bores, is one kind of teaching. Here is that extraordinary biologist, J.B.S. Haldane, who was never a bore:

> I consider it desirable that a man's or a woman's major research should be in a subject in which he or she has *not* taken a degree. To get a degree one has to learn a lot of facts and theories in a somewhat parrot-like manner.

This process, he says, makes it difficult to be highly original. Today, perhaps, Haldane's complaint is not always valid: the teaching of science is slowly changing—both in the schools and in the universities. Students are beginning to learn to investigate the unknown and to make independent discoveries.

Curiosity, discovery and finding facts are not the whole story. Students are often attracted to science because it gives power: to prevent disease (or to cause it); to produce more milk and more gadgets; and so on. The case history below is an example. So we now come to the third portrait: the scientist as technician—one might say, a *mere* technician. This idea has been popular among people in literary subjects. An example was a nineteenth century philosopher, G. Lowes Dickinson. He was notably gentle and kind but he was also an intellectual snob. He regarded scientists as no more than collectors of facts. So he classified the sciences with trade.

In fact, however, applied science can be just as exciting as the 'pure' variety. Often, it is also more difficult. Applying science may require attention not only to science but also to economics, politics, prejudice and much else.

PLAGUE IN WARTIME: A CASE HISTORY OF APPLIED SCIENCE

In 1945, during the late stages of the second world war, plague broke out in several Mediterranean countries. One was Malta, just emerging from a siege imposed by the aircraft and navies of Germany and Italy. Plague, especially without antibiotics, is more frightening than most human enemies. The carriers of the plague bacillus are rats, other rodents and their fleas.

I was sent out, a young zoology graduate, from the British scientific civil service, to advise on how to cope with the carriers. Behind me, I had the silent support and prestige of decades of dedicated scientific work by others. The rats and fleas had been classified and could be exactly identified. The facts of infection by a distinct microorganism had, by that time, long been an elementary part of medical science. In addition, and crucially, a group of zoologists in Oxford had recently been studying the behaviour of wild rats and their responses to poisons. They were helped by lack of experience of research on behaviour: their ignorance enabled them to make original observations on a much studied species.

(I had not, myself, taken much part in this work. I did, however, experience a sample of the odd demands of research when I spent one snowy Christmas Day in a barn near Oxford, recording the behaviour of animals in which my academic friends were interested. Despite the cold, the experience was not disagreeable. I learned something about animal behaviour which I later used in my own research. Also, I had an amiable companion. I think she later married one of the Oxford zoologists.)

When I arrived in Malta, I had, however, no high-falutin' thoughts about the power of the scientific knowledge and traditions on which I depended. My first problem was to convince hard pressed, sceptical medical officers that I knew what I was talking about. Then I had to contend with giving the same seminars four separate times to hygiene units from the army, navy, air force and the civil authority; to cope with the effects of immunization against plague (at that time, very disagreeable); and to keep up my fluid intake in hot weather when all the 'fresh' water was brackish. (Malta has no rivers, streams or lakes.)

The new, carefully worked out methods I recommended used much labour. It was first necessary to make a detailed survey, to find out what the very numerous rats were doing and to identify the species. Then an elaborate system of baiting was used, to train the rats to eat the bait. After that, a carefully chosen poison was offered. Then the baiting and poisoning were repeated, with different materials. This method had the crucial property of forcing rat numbers, if not to zero, at least to a point from which they could not quickly recover.

At any rate, the four sanitary squads were mobilized. The latest methods were applied. And the plague stopped. So did another disease carried by rats: endemic typhus, which had been smouldering in Malta for decades. So science had been applied in an emergency affecting about a quarter of a million people. No headlines resulted, nor (in 1945) were they expected or appropriate. I merely reported my experiences and conclusions, as a case history in public health, in the *Journal of Hygiene*.

Throughout the world, uncounted similar projects were and are silently in progress.

***

FACT

None of this gives a portrait of 'the scientist'. Is it possible to say what scientists are really like? A poet, Louis Macneice, has tried.

> A little dapper man but with shiny elbows
> And short keen sight, he lived by measuring things
> And died like a recurring decimal
> Run off the page, refusing to be curtailed;
> Died as they say in harness, still believing
> In science, reason, progress.

'Dapper', in the poem, is wrong. But many of us do believe in reason and progress. (In the absence of a well designed survey, I do not know how many.) We also think that what we do is worth while, for its own sake. Occasionally, somebody breaks silence and says so. One was Nikolaas Tinbergen (1907–1988), the principal founder of the modern science of animal behaviour.

> It is, I think, natural for a man to have occasional doubts about the value of what he is doing; at any rate, such doubts have often occurred to me. I find studying the behaviour of animals in their natural surroundings fascinating . . . it gives free scope to one's urge to observe and to reflect; and it leads to discoveries. Yet once in a while the embarrassing question comes up: 'So what?' [But] one usually ends by . . . concluding that it has all been worth while. It seems to me that no one need be ashamed of being curious about nature.

This is a muted comment; yet not many scientists would say even that, out loud. They might, however, be willing to acknowledge the need to be patient in exacting endeavours; to face frequent disappointment; to put up with incessant criticism; and to exert self criticism.

But no single image of *the* scientist is appropriate: the 'typical scientist' does not exist. To reinforce that, I quote Peter Medawar (1915–1987), a very exceptional scientist, who made a contribution to immunology of major importance for theory and for medicine. He was highly skilled and ingenious at the bench and he also had a profound understanding of theoretical biology and of the philosophy of science. I do not know where his successors are today. Here he is.

> Scientists are people of very dissimilar temperaments doing different things in very different ways. Among scientists are collectors, classifiers and compulsive tidiers-up; many are detectives by temperament and many are explorers. There are poet-scientists and philosopher-scientists . . . Most people who are in fact scientists could easily have been something else.

Scientists are not magicians, though they sometimes produce results that seem magical. They are not usually bores, though some teach science in a boring way. Most of them work with their hands and enjoy it, but much of what they do is different from a technician's work. Although the idea of science as exploration is sometimes dismissed as romantic, in fact even humdrum research can reveal the new and unexpected; and this can be exciting.

So science is a set of activities, done by a crowd of people, all different. But what is it? Are we justified in speaking of science as a single phenomenon?

PART IV

EXISTENCE

All the first seven chapters contain references to the uses and methods of science. But they nowhere commit the author to any clear statement about what science is or, indeed, whether 'science' refers to anything distinct. In the rest of the book I try to improve on this.

We begin with reduction, a topic which sometimes causes much agitation. Is science the doctrine that we are really *nothing but* animals or genes or atoms or ultimate particles? Some people hold explanatory reduction to be the scientific key to understanding both the universe and the human condition; others denounce it as a distraction from the study of whole organisms and a source of calamitous misunderstanding of humanity. Which is correct?

With reduction in its place, we return to language and to the logical and mathematical foundations of science. Like scientists, even mathematicians are found to be people: their subject, with science itself, is a growing outcome of human exertions.

It then becomes possible to outline the forms of human action called scientific. We ask whether there exists a logic of scientific discovery; whether there are moral foundations in what scientists do; and in what direction should science and scientists travel in the twenty-first century.

CHAPTER 8

ARE WE NOTHING BUT . . .? AND, IF SO, WHAT?

It may well be that the whole structure of physics is inadequate and misleading when extended too far into the processes of living matter.

DAVID BOHM, physicist

WHEN POPULAR FICTION DESCRIBES a future scientific age, it is often not a utopia but a distopian nightmare—something we should struggle to avoid. Some people find all science repellent, both the hardware and the software; for, in much of science, all we see and feel is broken down or *reduced* to imperceptible units, such as genes or atoms. Science thus seems to dismiss personal, sensual experience: all we enjoy; everything we suffer.

In 1820, the poet, John Keats (1795–1821), wrote famous lines protesting against 'philosophy', by which he meant natural science. It ruins the glorious rainbow, he says.

Do not all charms fly
At the mere touch of cold philosophy?
There was an awful rainbow once in heaven:
We know her woof, her texture; she is given
In the dull catalogue of common things.

Later, Edgar Allan Poe (1809–1849), the American poet famous for his horror stories, wrote a sonnet in which Science is harangued as a 'vulture, whose wings are dull realities'. In the same way, a modern poet, Cecil Day Lewis, imagines a scientific world and finds it repulsive.

121

Pasteurise mother's milk,
Spoon out the waters of comfort in kilograms,
Let love be clinic, let creation's pulse
Keep Greenwich time, guard creature
Against creator, and breed your supermen!
But not from me.

Day Lewis should have known better. By his time, meticulous scientific studies of the composition and other features of human milk had greatly benefited both mothers and infants. But, in 'reducing' milk to chemistry, scientists had not arrogantly trodden on the feelings of love and anxiety of mothers, still less those of babies. On the contrary, many must have been moved, as they still are, by empathy for both mothers and sucklings. (Among the true enemies of motherhood were the prudes who regarded breastfeeding as disgusting.) Similarly, if—like Hamlet—we wish to describe somebody as having 'Tears in his eyes, distraction in's aspect', we need not let ourselves be put off by the excellent recent research on the chemistry and physiology of the lachrymal glands.

Admittedly, a problem remains. The scientific and the sensual kinds of experience are separate: they seem unconnected, which is disconcerting. One absorbing kind of private experience is given by music. But music and its effects are, up to a point, also observable and repeatable; and repeatable phenomena can be studied scientifically (compare Chapter 1). In 1993, the science journal *Nature* published careful experiments, by F.H. Rauschner and others, in which hearing Mozart's Sonata in D Major for Two Pianos (K448) briefly improved the intellectual powers (in particular, spatio-temporal abilities) of undergraduates.

Perhaps that new finding will one day be expressed in terms of nerve physiology. Nonetheless, we can now, and shall in the future, still be able to appreciate what we have already enjoyed. For science does not say that enjoying a rainbow or a sonata or a lover is *nothing but* optics; or hormones; or genes; or anything else. Such a statement, which has been called 'nothing buttery', is not a scientific finding; and it is not clear why anyone should think that it is. It is not even sensible.

Nonetheless, it is important. So what can we say sensibly about reduction?

A sketch in black and white by a master of the fine arts, Rembrandt (Harmenz van Rijn, 1606–1669). Can it be reduced to and explained by the individual pen strokes of which it is composed?

THE REDUCTIONIST IMPERATIVE

In explanatory reduction, ideas taken from one kind of knowledge are used to explain the findings from another kind. Previous chapters contain many examples, some of them misguided. Human character has been reduced to cranial bumps; friendship, enmity and war to animal social life; our social conduct to genes.

But such lapses from common sense by no means oblige us to reject reduction. Its clearest successes have been in the physical sciences. The apparently infinite number of substances around us have been reduced to about a hundred elements and the elements to a smaller number of ultimate particles. As a result, in less than two centuries our knowledge both of the inanimate world and also of organisms has been transformed. In biology, the leading case is explaining heredity by cell chemistry (chapter 5).

One consequence has been the disappearance of vitalism. For millennia the strange properties of living things had been attributed to occult inner qualities. Residues, such as drive and instinct (chapters 3 and 4), still appear in biological writings. Similarly, until the seventeenth century, our constant body temperature was attributed to an undefined 'innate heat'. Today, physiologists have given this up. They have exact (and useful) knowledge of the chemical changes which produce the heat: body heat is reduced to chemistry.

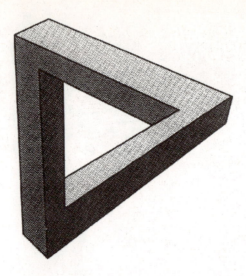

Limitation of reduction illustrated by an impossible triangle. Can the impossibility be *reduced* to any one part of the figure?

Two outstanding biochemists are among the many who have therefore espoused an uncompromising reduction: they have seemed to say that a kind of knowledge exists, superior to and capable of replacing all others. Their doctrine becomes reductionism (or scientism). Francis Crick has written that 'the ultimate aim of the modern movement in biology is . . . to explain all biology in terms of physics and chemistry'. And Jacques Monod (1910–1976), in a much quoted popular work, states that 'the ultimate aim of the whole of science . . . is to clarify man's relationship to the universe'; and that the baffling phenomena of biological science, including those concerning humanity, can be unravelled only by 'genetic and biochemical analysis'.

The researches of Crick, Monod and their colleagues were based on reducing living things to their chemistry and have helped to found modern biology. It is therefore easy to believe that all good science consists of reduction. In 1980, I asked 61 advanced students of zoology to comment on this sentence: *All biological phenomena can, in the long run, be explained in terms of the physical sciences*. With two colleagues, I analyzed their answers. More than half the students (and several of their teachers) accepted this reductionist proposition as obviously true. Yet physical science is

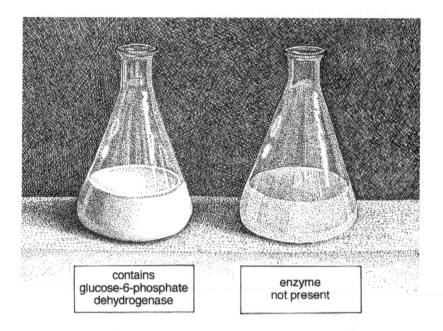

contains
glucose-6-phosphate
dehydrogenase

enzyme
not present

Reductionist statements which might be made by a biochemist. Each flask contains fluid from human red blood cells. What questions come to the reader's mind on seeing them?

not putting biologists out of business; and to imply that biologists will eventually be excused from studying whole organisms is not a scientific finding but a metaphysical presumption. It is also, as we find below, obviously wrong.

The limitations of reduction are not confined to the living world. Roger Penrose provides an ingenious instance from geometry. In an article, 'Must mathematical physics be reductionist?', he reproduces the famous drawing of an impossible wood triangle shown on page 124; and he asks, 'where is the impossibility?' If any one corner of the figure is covered, what remains could represent a real object. The impossibility cannot be reduced to a single part: it is evident only in the whole. He gives other geometrical constructions with features which are properties of wholes. He could have added works of art to his examples. The impact of Jan Vermeer's famous 'View of Delft' comes from the whole picture, not from its separate details or from the chemistry of the paint.

For more limitations of reduction look at the figure above,

in which biochemical statements are made about two fluids. What questions come to mind on seeing this picture? Some people might ask about the chemistry of the fluids—a reductionist question; but I suspect that most would want to know about the people from whom the blood was taken.

The enzyme G–6–PD is usually present in human blood but is deficient in the blood of many Africans. The defect, which results in early death, is due to the presence of two copies of a usually rare gene. Yet in some malarial regions the deficiency is common. Why has it not been 'weeded out' by natural selection? The people with only one copy of the gene are resistant to malaria, hence are biologically fitter than those with two copies of the gene or with none. For populations in malarial regions, it is therefore advantageous to have a high incidence of the gene. When malaria is finally eradicated, the gene will slowly disappear. This explanation is not reductionist: it begins with observation of people and depends on knowing their environmental relationships (their ecology). As so often, *a full understanding requires both reductionist and nonreductionist information.*

But even this example is not enough. Consider a community in which many people are weakened by a deficiency disease, beri-beri. A biochemist recommends food with extra vitamin B_1 but is dismayed when this rational policy is not adopted: the new diet is rejected as disgusting, improper or unsuitable for cooking. In a strange culture, a reductionist policy, however well founded, may be ineffectual unless it allows for longstanding customs. These can rarely be changed by manipulating people as if they were chemical systems or animals. Reduction is necessary; but human beings must also be respected as persons living in a social environment.

REDUCTIONISM AND LOGIC

Suppose that we are urged to disregard common sense and to agree that all biology is reducible to physics: that is, that the laws of physics, concerning the movements and interactions of atoms and ultimate particles, can tell us everything that there is to know about all larger systems—molecules, organisms, societies . . . These, of course, include human beings and the things they say, such as statements about reduction.

We may legitimately reply: to say that everything can be reduced to physics is not testable by observation or experiment. It is not part of physics or any other science: it comes in from a different realm of knowledge. The following statements belong in that realm. *The existence of physics itself rules out an uncompromising reduction: physics is a branch of knowledge; and knowledge implies the existence of a knower.*

Similarly, when people make statements, they use language according to sets of rules; and their assertions are sometimes based on arguments which follow logical principles. The noises people make when speaking can be analyzed by the principles of acoustics; but the rules which govern language and logical argument cannot be derived from the physical sciences.

This conclusion can be reinforced by examining not reduction but emergence. One doctrine of reduction implies that, to understand an organism, we need to know only the atoms of which it is composed. But the individuality of DNA emerges from the relationships between its parts. Its individual features cannot be predicted from a list of chemical components.

This principle is of extraordinary importance for the understanding of human differences. Our distinctive traits emerge, during individual development, from interactions with a series of environments. Similarly, social phenomena, such as property ownership, health services, schools and orchestras, emerge from relationships between persons.

Emphasis on emergence does not, however, reject reduction as a method. When it is successful, reduction often explains the large by the small. Chemical facts are needed for a nutritional policy and information on microorganisms for a system of public health.

We therefore have a sequence or 'hierarchy' of kinds of understanding. Overarching all other kinds of knowledge is our awareness of ourselves and our history (chapter 6). Within objective studies, at one level or tier are the ideas and phenomena of the social sciences: oligarchy and democracy; the gross national product; the nuclear family; education . . . When we move into science, we find none of these in biological texts: there instead we may find respiration, territory, gene and statements about survival value. None of these, in turn, is part of physical science. And the physical sciences themselves offer a hierarchy, from nebulae and black holes to the mysteries of ultimate particles.

AT THE GRASS ROOTS

What are the implications of all this for scientific research? Here is an imagined, down to earth scene, to give an impression of what scientific action is like. (A real example, involving reduction, is in Appendix 2.)

A botanist has to study an unfamiliar species of plant. She is very bright and skilful, but she is an ordinary scientist, not a genius. (Nor is she rich.) She is fortunate because her work is satisfying and she is a valued member of her community; but she is unlikely to appear in the media. (Just to make the portrait less skeletal, she is in her forties, plays tennis and goes bushwalking; enjoys music from Bach to Bruch; and reads modern novels. Her husband is a cheerful secondary school teacher of History and English; and they have two children, both with the usual teenage problems.) She also knows this:

> . . . scientific research is very difficult. Anything that can go wrong will go wrong. No one, no matter how experienced, can do a complex experiment without the guidance and criticism of others. Isolation is the death of science.

That was written by a physicist, Alan Cromer; but it applies to all science.

Our botanist has seeds from the plant she is to study. What she does with them depends on the questions that interest her. She might begin by analyzing them chemically and so discover their nutritional value; or she might grow some and find a new, useful drug, like digitoxin, quinine or morphine. Or she could study their DNA and so find gene sequences which enlarge the plant's productivity or enable it to resist disease. *These are examples of reduction*. The plants are, so to speak, reduced to chemistry or to genes. The method is immensely powerful; but the botanist must still begin with knowledge of the plants themselves.

Some possible projects would be on whole plants. The botanist and her colleagues might set the seeds in good soil, observe the resulting growth and so estimate the yield of food. They might also set samples in a variety of soils, and decide which soil is the most favourable. Or they could analyze the variation within and between populations, to find out how much is genetically caused and how much is environmental; which, as we know, is very important but not easy. They could also find out whether the plants would grow well and look attractive in gardens.

Only one or two of those projects could be attempted. Each would be almost entirely about details. The botanist might form a hypothesis about the influence on growth of a trace element in the soil; and it could take her years to test it. The hypothesis would not have originated by mutation from another notion or been subjected to natural selection. It might well, however, have arisen nonlogically. The researchers would then probably have to design, or at least learn to use, elaborate apparatus. And as the results began to come in they would be given logical criticism and statistical analysis.

All the projects imagined above have possible economic uses. They therefore leave out an important fact about the botanist: she and her colleagues are dedicated to finding out more about plant life for its own sake. This is true even if they are required to pay attention to profits. Or, if they are in a university department, they may be struggling with diminished funds. They are then also teaching botany to undergraduates and research methods to graduate students. And they enjoy all this (except their inadequate resources).

Now back to reduction. Suppose that the botanist concentrates on engineering genes and so produces a Fur Tree: a new species, *Arbor villosissimus*, which means a very hairy tree—something from which you could make a fur coat. It may be superbly engineered but it needs an environment to grow in: it still requires soil and water and carbon dioxide. So, to cultivate it, she would have to study the whole organism in relation to its surroundings. She would also have to find people to grow it and to finance her research; and more.

The critical reader may now protest that an organism, even a human being, can be broken down to chemical elements; and, when that is done, it is found to consist only of those elements: no 'vital essence' wafts out, like a medium's ectoplasm. This is true (compare the comment on vitalism above). But, when atoms, compounds, genes or organisms are rearranged, new things emerge. The outcome of combining items cannot be plucked from knowledge of the separate parts. Chemistry and physics may greatly help the study of the new wholes but it cannot replace them.

> Scientists *have* to welcome reductionism as a *method*. But, before we can even attempt reduction, we need as great and as detailed a knowledge as possible of . . . what we are trying to reduce. Thus before we can attempt a reduction, we need to work on the level of the thing to be reduced (that is, on the level of 'wholes').

That common sense we owe to the philosopher, Karl Popper. The fundamental principle is this: explaining by reduction does not, and cannot, do away with what is explained.

WILLIAM OF OCKHAM'S BLUNT INSTRUMENT

Devotees of reductionism often quote with approval a saying by an English theologian, William of Ockham (1300–1347?), who was excommunicated for criticizing the Pope and for advocating poverty among Christian priests. His principle, put simply, was that explanations should always be as simple as possible. (He wrote: *pluritalitas non est ponenda sine necessitate*, but is often misquoted.)

Although 'Ockham's razor' is sometimes said to be a scientific procedure, it is neither a scientific finding nor a logical principle. It is a metaphysical or psychological directive concerning scientific explanations: it tells scientists what they *ought* to do. In a famous work, the Austrian philosopher, Ludwig Wittgenstein (1889–1951), wrote:

> . . . the process of assuming the simplest law has no logical foundation but only a psychological one. It is clear that there are no grounds for believing that the simplest course of events will really happen.

The bluntness of this ill named razor has, since then, been brought out in studies of humanity. To explain what human beings do, we must often go from simple to complex. As I show earlier, a nutritionist who tries to reduce people to nothing but chemical systems or animals is likely to be ineffective. The refusal to recognise complexity has also led to the errors of instinctivism (chapter 4) and to the attempts, by sociobiologists and others, to reduce human beings to animals or bags of genes (chapters 3 and 6).

PERSONS

To see more of the limitations of explanatory reduction, we might begin by attempting a strict reductionist's account of a passionate kiss. This would require, with much else, an accurate knowledge of the muscles of the face, such as the very complicated *orbicularis oris* around the mouth. I leave that attempt to the reader. Here, instead, is a reductionist's all-embracing statement about the human species.

> The human animal is a chemical machine and the brain is a computer. It is also a product of evolution by natural selection. What human beings do is fixed by their genes. Therefore, human society can be explained only by resort to biology and chemistry.

Earlier I quote leading scientists who have made assertions something like this. They seem to imply that we are nothing but assemblies—of what? genes? elements? ultimate particles? We, *and* the people who make these statements.

Such assertions have spread into the media. A much admired popularizer, Richard Dawkins, made a mark in 1976 by saying, in *The Selfish Gene*, that we, human beings, are machines created by our genes and that we are therefore born egoists. And, presumably, that he is too. This hardly leaves scope for moral teaching. Twenty years later, he turned to describing people as self duplicating robots. He also provides a definition:

> A robot is any mechanism, of unspecified complexity and intelligence, which is set up in advance to work towards fulfilling a particular task.

Here is a muddle, one typical of such writings. A machine or mechanism is indeed something made 'in advance' by people; and it has a function such as pumping water, telling the time or doing sums. But a person is not an item of equipment manufactured from a designer's drawings. Nor is an animal or a plant or even a bacterium. Similarly, the human brain is not a computer (chapter 2). A computer's performance depends on the programs chosen by its maker or owner. Its 'memory' can be wiped clean and given new programs. This does not apply to the reader's (or the reductionist's) brain.

In a book on evolution, where organisms, including people,

are called robots, the author refers to a famous computer scientist, Alan Turing. He was, we are told,

> the young British mathematician who, through his codebreaking genius, may have done more than any other on the Allied side to win the Second World War, but who was driven to suicide after the war by judicial persecution, including enforced hormone injections, for his homosexuality.

This writer is describing a tragic response to maltreatment and is expressing indignation at an outrage. The writer himself comes over, not as a computerized robot, but as a humane person, with 'affections, senses, passions'. This author, as you have no doubt realised, is Richard Dawkins.

So popular reductionists disregard obvious facts about themselves and about human social life. When they do this, they also confuse their readers about science, especially biology—its scope, and its limitations.

CONSCIOUSNESS AND PAIN

Suppose a physician or surgeon is faced with a person brought in after an accident. An early question is whether the patient is conscious. In this case, 'conscious' means 'responsive': does the patient answer questions—know her name, the date and the capital city of her country? These questions, at first sight, seem to *reduce* the patient to a system which can be studied impersonally, from outside. (They are 'behaviouristic'.)

Next, the patient (if responsive) may be asked how she feels; is she in pain? This question, too, could be regarded as behaviouristic: it is designed to provoke an answer such as, 'My leg hurts'. But most physicians, like the rest of us, would regard it as more than that: the patient is now treated as a person with feelings—such as pain; and emotions—such as anxiety. These can be directly observed only by the patient herself. Others have to infer their existence.

Pain, however, can also be usefully studied objectively: referred pain, for instance. A phenomenon known to many of us is sciatica, a pain in the leg. It is usually due, not to anything in the leg, but to pressure on a nerve from a prolapsed disc in the back. Knowing this makes effective treatment possible. Moreover, a surgeon *can* operate effectively on the disc without knowing anything about the patient as a person. (But this is unusual.)

Then there is the patient herself, who is, most of the time, not only conscious but self conscious. The importance of such knowledge has long been recognized but, in a period of behaviorism and robotics, can easily be ignored. The Latin *conscientia* and the Greek equivalent, συνέιδησισ, both referred to the knowledge one has *of oneself*. We all possess this knowledge. The patient may be telling herself that she should not complain without need and that she must be helpful to the doctors and nurses around her.

These brief paragraphs move between reductionist statements, such as those about the mechanics of the vertebral column, and references to the patient as a person with feelings and intentions. In the previous paragraph she is aware of herself as a person with a conscience. A longer narrative would include the ethical principles which influence also the conduct of the doctors and nurses. Consciousness, conscience and self consciousness cannot be understood if they are reduced to biology or physics.

Some scientists assume reductionism without giving it much thought, others reject it. Here is a neurologist, G.M. Edelman:

> A person is not explainable in . . . physiological terms alone. To reduce a theory of an individual's behaviour to a theory of molecular interactions is simply silly.

Why is it silly? Earlier pages partly answer that, because they bring out the significance of persons. The remark by Edelman comes from a meeting where reduction was discussed. A psychologist, Margaret Boden, spoke on 'Artificial Intelligence and Human Dignity'. She quotes resentful remarks made by American workers during a study of attitudes to their work. She writes:

> These workers took for granted, as most people do, that there is a clear distinction between humans on the one hand and animals—and machines—on the other. They took for granted, too, that this distinction is grounded in the variety of human skills and, above all, in personal autonomy. When their working conditions gave no scope for their skills and autonomy, they experienced not merely frustration but also personal threat.

The disregard of persons, in writings offered as science, arises from the attempt to reduce human beings to chemical or electrical systems or to animals. Such writings, in which people are sometimes described as robots, are fraudulent: they offer non-science as science; they obstruct understanding of what we know; and they present preposterous caricatures of humanity. Correspondingly, the previous chapter offers, as one portrait of 'the scientist', a cold, detached and above all logical thinker uninterested in persons: never the lunatic, the lover or the poet whom Shakespeare brings before us in *A Midsummer Night's Dream*. This legend too is wrong.

Yet science does demand logical thought. So, in the next chapters, the thoughtful reader is asked to think about thinking.

CHAPTER 9

METAPHOR:
A BRIDGE PASSAGE

Long ago I spoke of . . . a hazy metaphorical contrast between warm mammalians who tenderly suckle their living creations and cold reptilian intellectuals who lay abstract eggs.

NORTHROP FRYE

WHEN SCIENTISTS USE EXPLANATORY reduction, they move ideas from one realm to another; and often, at the same time, they resort to metaphor: they transfer, not ideas or concepts, but *words* to a new use. Such transfers are a universal practice. The present short chapter is designed to remind the reader of this central feature of human intelligence and the use of language.

Metaphor, however, at first sight seems to be the territory of poets and other literary persons. Here is a matter of fact ('scientific') statement about a dawn.

The sun rose at 05.23 hours; cloud cover was four tenths.

But John Milton describes a similar scene,

> . . . when the sun in bed,
> Curtain'd with cloudy red,
> Pillows his chin upon an orient wave . . .

and makes the rising sun a person. And Shakespeare provides different images to show us a glorious morning,

> Kissing with golden face the meadows green,
> Gilding pale streams with heavenly alchemy.

After these fragments, ruthlessly torn from two of England's greatest poets, the reader may expect a harangue on the contrast of poetic imagery with scientific reports and an exhortation to scientists to leave fancy writing to others. Writing in plain words is indeed desirable (and a struggle to achieve); but to urge scientists, or indeed anyone else, to discard all metaphor would be both stultifying and pointless: stultifying, because metaphor greatly helps our communication and our thinking; pointless, because metaphor is an unavoidable feature of language. It is often hidden; it is a powerful means of persuasion; it is an essential component of creative thinking; and it is fun. Indeed, all language is itself, in a sense, a series of metaphors: it uses words to refer to things and thoughts.

CAPRICE

The reader, faced with a whole chapter on metaphor, may nonetheless accuse me of being capricious in choosing to write it and so compare me with a she goat (*capra*). And indeed, I vacillated (*vaca*, a cow) before embarking on it.

Despite the metaphors, the sentences in the previous paragraph are descriptive. The same applies to the expression, bridge passage, in the chapter title: there, it is a metaphor piled on a metaphor: in music, 'bridge' is used to refer to a long and elaborate sequence which links two distinct themes. All such terms are examples of buried metaphor. Each may be regarded as a play on words, in which we take part usually without knowing it.

PERSUASION

Description is often far from plain: it may be carefully propagandist. The speaker or writer may know that the attitude of an audience to a thing or an idea can depend on just how it is described. Imagery and metaphor may be used to persuade us that a statement is correct. In a famous and controversial treatise, *A Study of History*, A.J. Toynbee (1889–1975) describes the Spartans of ancient Greece, the nomadic people of the Russian steppe and the Eskimo as examples of arrested development. Each is said to have a system of castes with strict division of labour between them. Their social systems are held to be responsible for their failure to make any social progress. Toynbee then suggests that 'corroborative evidence' is provided by creatures such as ants and bees: these insects, he says, have the same features, especially castes, as arrested civilizations.

But, as we know, biological evolution and human social evolution are quite different. The social systems of insects do indeed remain the same, generation after generation: detectable change probably occurs only after millennia. As well, the behaviour of each species is uniform, with only minor variations. Human social action, including that of the Spartans and the others, has the opposite features. Hence identifying human societies with those of ants or bees cannot give rational support to an argument about human history.

Yet such metaphors and analogies are rife, notably in political science. Writers in this field obviously enjoy them. Some, however, have complained. G.D.H. Cole (1889–1959), economist and historian, once urged his colleagues to give up likening society to a mechanism, an animal or a person and to use a more appropriate terminology.

I doubt whether many of them took any notice. Chapters 3 and 6 provide many examples, some very recent, of likening people to animals to support social attitudes. Among them are equating human property with animal territories, human sex life with that of birds and lying with the deceptive appearances of insects. Similarly, for more than a century both right and left have resorted to Darwinism to prop up political theories. Metaphors from garbled biology can be used to advertise 'ideologies' of any kind whatever.

METAPHOR UNAVOIDABLE: SCIENTIFIC 'LAWS' AND IMAGERY

Yet metaphor is relevant to real science. Consider the expression, 'the laws of nature'. The *Oxford English Dictionary*, under the heading 'human law', gives, as the primary meaning of 'laws', a body of rules, formally enacted or customary, recognised as binding on the members of a community. Laws in this sense vary from one community to another and from time to time:

they are products of human argument, reason, unreason and action.

But, in the context of science, the word 'law' refers to a regularity discovered in nature. Such laws are inescapable. This usage has been called a 'grotesque pun'. Unlike a law in the first sense, a 'law of nature' cannot be rescinded: we have to put up with it. We may, however, be able to apply it, as when Ohm's law, concerning the flow of electric current, helps engineers to supply us with electricity. Or we may get round a principle, such as that of gravitation, as when we make machines which enable us to fly.

But this second usage has sometimes led to more verbal manipulation and so to absurd errors. If the regularities observed by scientists are indeed 'laws' then—according to some thinkers—perhaps after all they do resemble laws in the first sense. In which case, there must be a Lawgiver. (Compare the Absolute in chapter 6.) Such an argument disregards the two entirely different meanings of 'law'.

All this may suggest still more of a gulf between authentic scientific narrative, on one side, and literary or religious discourse, on the other. But, if it does that, it is misleading, for the writings of great scientists are riddled with metaphor. (For sources, see my *Biology and Freedom*, chapter 2.) When William Harvey (1578–1657) announced his fundamental findings on the circulation of the blood, he identified the heart first with a Prince because all the organs of the body depend on it. But later he likened it to the sun, because the heat of the sun promotes the circulation of water through the earth and the air. Such literary flights can help the writer and, often, also the reader.

In the nineteenth century, leading physicists happily described heat as a *fluid*. Michael Faraday (1791–1867) wrote of *elastic* fluids and proposed a *field* theory. Their findings stemmed in part from the much earlier work of Johannes Kepler (1571–1630), who had presented the universe as a system like a clock. This was, in its implications, more than a literary device, for clocks are mechanisms to tinker with. Hence Kepler's metaphor goes with an attitude to nature as something which can come under human control and be managed. It matches the concept of science as knowledge that confers power.

Metaphor helps scientists not only to describe their findings but also to make conjectures. Darwin's crucial concept of natural selection relates evolutionary change in nature to the *selection* by farmers of seeds or stock for breeding. Famously, Darwin was also influenced by the writings of a pioneering demographer, the English clergyman, Thomas Malthus (1766–1834), on the ways in which human populations are regulated. Malthus held human numbers to be always in danger of exceeding the resources needed for survival. Hence followed the slogans of a struggle for existence and of a *war* of nature. Darwin, we know, used other metaphors lavishly from banking, business and industry. These helped both Darwin and his readers; but, alas, they also encouraged people to portray humanity, by analogy, as compulsively violent and competitive (and many still do so).

THE WAY WE THINK

Metaphor, as we now see, pervades educated speech and writing, yet it is nonlogical. It reflects the highly developed human capacity to perceive resemblances intuitively and to draw conclusions from them.

In *The Science of Life*, I call this skill the Marple Principle, after Agatha Christie's celebrated detective. Jane Marple's achievements are based on her ability to observe likenesses. The people she meets in her encounters with crime are like those she has known in her English village. Her large acquaintance with ordinary people enables her to identify criminals. She works intuitively ('by instinct'). She exemplifies common sense—a range of abilities which we all possess and use every day.

These abilities are shown in our mysterious capacity to recognize objects and to associate one thing with another. If I now write 'miaow', that will, without delay or logical analysis, call up before the reader the word cat and the image of a familiar mammal. It may even evoke a memory of the kind of behaviour called catty.

In an age dominated by machines, especially computers, we are likely to ask: by what strange *mechanisms* are such responses produced? An American psychologist, Jeremy Campbell, has indeed put out an excellent book on the brain and intelligence, entitled *The Improbable Machine*. Despite his title, Campbell's leading theme is the contrast between human thinking and the operations of even the most advanced computers (compare page 32).

Computers are logical. They use the propositions of formal logic and of mathematics. These rarely contain ambiguities: each term is precisely defined. The words of our everyday speech are quite different. According to Campbell, English words such as 'set' and 'run' each have about 140 meanings or uses. I have not counted them; but I can illustrate the complexities of English, and indeed of language generally, from the problems of machine translation.

A computer scientist is said to have used advanced equipment to translate several familiar proverbial sayings from English into Japanese and back again. After this double process, one emerged as: *blind men are mad*. What was the English original? The reader may care to pause and think (nonlogically) about it. The answer is not obvious; but, when arrived at, it does illustrate the inability of a machine to ask, from general knowledge, 'does this make sense?' To ask such a question requires an understanding (intuitive in the reader) of the importance of context and of the fact that a single word has different meanings, sometimes metaphorical, in different sentences. In case you have not already worked it out, the answer is: *out of sight, out of mind*.

In our perception of the world and of words, we interpret what we see and hear. We classify and match, or mismatch, one thing with another. We rarely make formal, logical deductions. Science, however, employs both intuition and logic; and it often depends on the kind of logic called mathematics.

Yet, despite the need for formal logic and the hazards of metaphor, I now ask the reader to allow me a little frivolity. For I feel *licensed* to describe my present writing as an *unfinished journey*, perhaps an *uphill climb*, toward an understanding both of science and also of what we can learn from science about ourselves. Is there, we may ask, a kind of *class struggle* between Miss Marple and the logicians? Are matching and metaphor *enemies* of objective, scientific knowledge? Are scientists after all, in Northrop Frye's words, cold reptilian intellectuals who lay abstract eggs? Read on.

CHAPTER 10

SCIENCE AND SUMS

It must have taken many ages to discover that a brace of pheasants and a couple of days are both instances of the number two: the degree of abstraction involved is far from easy.

BERTRAND RUSSELL

MUCH SCIENTIFIC KNOWLEDGE CONSISTS of abstractions, especially those that involve mathematics; but in science a first step usually consists of identifying distinct kinds of easily observed objects, such as species of plants or animals. This is essential but requires no elaborate logic. Compare the importance of classifying things (page 113). A likely next step, also at first descriptive, is to count, to measure or to weigh them. If the objects observed are the sun, moon and stars, measurements of angles may be needed, such as those used for founding astronomy, for making a calendar and for setting up early stone monuments. All this began before writing.

Later, but still in the remote past, came the need to calculate the areas of fields and how many bricks are needed to build a temple. These skills are now taken for granted. So is time keeping; but it took a long time for mechanical clocks to be invented. Before then, in the second century of our era, the great Greek encyclopedist, map maker and mathematician, Ptolemy, (Claudius Ptolemaeus of Alexandria, ?90–168 CE), worked out the quite tricky geometry needed for designing sundials.

Mathematics has been not only an essential component of science and technology but also a puzzle for philosophers and even an object of worship. The reader may wish to add: and a source of torture for uncounted schoolchildren. (This was acknowledged in a famous mathematician's names for the branches of arithmetic: 'ambition, distraction, uglification and derision'. The mathematician was Charles Dodgson, alias Lewis Carroll, 1832–1898.)

So in this chapter we again meet a product of human thought and action which, like language and music, is confined to humanity: not even chimpanzees do long division, let alone discuss the problems of probability. Like other skills, mathematics has to be learned. Much remains to be found out on how to teach it effectively. And, unfortunately, however good the teaching, the imposing edifice of advanced maths seems to be accessible to only a few. Worse, fragments of it are often misunderstood, even misused. In this it resembles evolutionary theory and genetics. Hence, in some parts of this chapter, I have again to try to clear away obstacles to understanding.

A reader who has not studied physical science, and wishes to see more of the history and massive positive contributions of mathematics, may turn to *Mathematics for the Million* by Lancelot Hogben (1895–1975). Hogben, a zoologist, was remarkable and so was his book. It was first published in the 1930s but is still in print. Hogben wrote it in hospital, during a long illness, for his own fun; hardly credible. He writes:

> the reasons which repel many people from studying mathematics are not at all discreditable. As it has been taught and expounded in schools no effort is made to show its social history, its significance in our own social lives, the immense dependence of civilized mankind upon it. Neither as children nor as adults are we told how the knowledge of this grammar has been used again and again throughout history to assist in the liberation of mankind from superstition.

But he also warns his readers against a 'blind reverence' for numbers.

A LONG STRAIGHT LINE FROM EUCLID

For many of us, an early experience of abstract, mathematical logic was 'Euclid'. Geometry was given this name after a Greek genius who lived around 330–275 BC. His work is well known but not his life. His 'complete' system of geometry, a dazzling achievement, remained almost unchanged and unchallenged for more than two millennia. It codifies the procedures used in such everyday affairs as measuring fields.

When children begin Euclid, they are commonly required to learn certain axioms, such as 'a straight line is the shortest distance between two points' and 'parallel lines never meet'. These are not proved: they are 'given' as self evident. From them can be derived a sequence of theorems, culminating in the famous one, usually attributed to Pythagoras, about the square on the hypotenuse.

The responses of children to geometry vary. Most, I suppose, accept it with some degree of resignation, especially if it is taught as a purely theoretical set of exercises. An exception was Bertrand Russell. When his elder brother wanted to teach him geometry, he immediately asked why he should accept the axioms if they could not be proved. Eventually, but only in order to continue, he agreed to assume the axioms provisionally. (He must have been a very irritating child.)

The young Russell raised, by implication, questions much debated by mathematicians. Later, in his *Introduction to Mathematical Philosophy*, he accepted the need for an unproved beginning. He wrote:

> Since all terms that are defined are defined by means of other terms, . . . human knowledge must be content to accept some terms as intelligible without definition, in order to have a starting point for its definitions.

The reader may find that statement intuitively obvious or at least easily acceptable. But not all mathematicians have done so: some have tried to create a logical system which could stand by itself, without the need for unproved assumptions. They were following a tradition, traceable back for two and a half millennia, to Pythagoras (?580–500 BCE). Pythagoras and his still more influential disciple, Plato, held that what we know is only an imperfect copy of a remote world of ideas: for them, mathematics is part of that world and exists, complete and perfect, outside space and time, perhaps in a divine mind like the Absolute of nineteenth century philosophers (page 83). It is then the only means of achieving Truth: 'real is what can be measured'. In the words of Leopold Kronecker (1823–1891), 'God made the integers; all else is the work of man'.

In 1932, however, a Czech mathematician, Kurt Gödel (1906–1978), finally disposed of this notion. He presented a formal proof, now generally accepted, that no self-contained mathematical or logical system can exist. Calculations are valid if they conform with certain rules; but the rules cannot themselves be validated by the same rules. As Russell had said, it is always necessary to begin a logical or mathematical system with something not proved. In showing this, Gödel confirmed the status of mathematics, not as a divine dispensation but, like science, as a product, still growing, of human intuition and reason.

Euclid Dethroned, Nothing Installed

Here are three famous incidents in the history of mathematics. Each illustrates the growth (not always logical) of mathematics as a product of human intelligence.

- The first, which is very simple indeed, concerns Euclid's straight line. Imagine four points on such a line, a, b, c and d; b is between a and c; c is between b and d. You will agree—without proof—that both b and c must be between a and d. Only in 1882 was it pointed out that this obvious conclusion cannot be reached from Euclid's system. In this case, intuitive understanding is ahead of logic.

- The second concerns the axiom about straight parallel lines: put formally, it states that, through any point p not on a given line L, there passes only one line parallel to L. In the nineteenth century, several mathematicians, rather like the young Russell, asked what would happen to geometry if this axiom were discarded. As a result, they produced new geometries in which parallel lines do meet, like the lines of longitude on the surface of a globe which are parallel only at the equator.

One outcome was the 'elliptical' space developed by an outstanding German geometer, G.F.B. Riemann (1826–1866). J.B.S. Haldane, in a collection of essays, *Possible Worlds*, describes how he imagined himself in such a space. He is standing on a transparent plane which he sees by looking down. If he looks up, he sees the other side of the plane and, through it, the soles of his feet pointing backwards.

This is not an elaborate mathematical joke about an impossible world. One of the foundations of modern physics is the general theory of relativity of Albert Einstein (1879–1955). In this theory, the cosmos is a nonEuclidean, curved space–time; and, to develop it, Einstein required Riemann's outlandish geometry. Here is one of the most impressive examples of the use of pure mathematics, a product of abstract thought, to solve problems concerning the material world.

In this case, formal logic is ahead of intuition. The existence of several geometries also allows scientists to choose whatever mathematical system they need to help them. No Absolute exists. The question is: which system is the most convenient?

- Last, we turn to nothing. J.R. Brannin has described how the difficult concept of zero sometimes confuses children of nine or ten. If they are required to subtract 120 from 235, and to begin by taking zero from 5, they may write 0 instead of 5.

They need not be embarrassed, for when, in the twelfth century, zero was introduced from India, through Persia, to Europe, it troubled many learned persons. Faced with such expressions as 10 and 100, they exclaimed, 'What! You add nothing to one and make it ten; and nothing again and make a hundred! What nonsense!'

Today, few of us, at least after the age of ten, have difficulty with zero. We learn to use it early enough to take it for granted. This, of course, is not instinctive knowledge: it is a logical process learned from a teacher. The ability of teachers to transmit this, at first difficult, idea to young children is owed to the gradual development of a longstanding tradition.

REELING, WRITHING AND . . .

The beginning of arithmetic, like that of language, cannot be studied. Again, what we can do is observe its development in children. We then meet, once more, that monster with many heads, the interaction of heredity and environment. Stanislas Dehaene describes how children have been supposed to be 'endowed with unlearned principles of counting'. They are said to invent counting for themselves: counting is described as 'innate'.

Such statements come from asking the wrong question,

namely, is counting innate or learned? Such a question cannot be answered. As usual, the proper questions to ask are: how does the skill develop? what factors influence its growth? how can we encourage it?

As J.R. Brannin shows, answers are being found by minute observation and testing of children. Very young children do indeed easily learn to count a group of objects. Later, they find that counting tells them *how many* objects are present. This and similar abilities are enhanced if they take the advice of the musician who was asked, by a visitor to Washington, how he could get to Carnegie Hall. The musician replied, 'Practice, practice!'.

The need for incessant practice is related to a feature of our thinking encountered earlier: the Marple Principle. We have a marvellous memory for diverse experiences; hence we can quickly understand and explain complex situations. Also, unlike a computer, we can instantly interpret a great variety of shapes. But a quite different kind of skill is needed quickly to solve (say) 73×249, let alone a problem in long division. Such calculations can be polished off by a computer in a few millionths of a second. Dehaene therefore calls even the multiplication table 'an unnatural practice'.

ACHIEVEMENTS

Unfortunately, however much we try, when we learn elementary maths we often learn, or fail to learn, only elementary maths. As Hogben wrote (above), we do not hear about its uses and social context. So we now turn from difficulties to glimpses of what mathematics has done for science.

A leading case was provided by the famous priest, J.G. Mendel. As every schoolchild knows, he made a long series of crosses of varieties of the garden pea (*Lathyrus*); and he *counted* the different kinds of offspring. From the observed ratios, he deduced the concept of biological heredity based on paired discrete factors (now called genes). At first, these factors were abstractions. Soon it became obvious that they must lurk in the chromosomes of the cell nucleus (page 77). They were eventually given full chemical reality long after Mendel's findings were rediscovered in 1900. Today, founded on Mendel's original ratios, we have the giant molecule, DNA, and the giant structure of modern genetics.

Similarly, the kinetic theory of gases is a foundation of much modern physics. It arises from the fact that gases combine to form new substances in constant numerical proportions. From this it was inferred that gases consist of discrete items (now called atoms or molecules—compare genes).

Another fundamental achievement was the periodic table of elements assembled by a Russian chemist, D.I. Mendeléev (1834–1907). If the chemical elements are arranged in the order of their atomic weights, their properties are found to vary in a regular way. Mendeléev based his table on the 63 elements which had been identified in 1869. (Now, more than one hundred are known.) The numerical findings allowed predictions concerning the elements needed to fill the gaps in the table. Mendeléev himself predicted the properties of three elements, today known as gallium, scandium and germanium.

The order is now based on atomic numbers, not weights; and later developments in the physical sciences have some likeness to those in genetics. For, in another great advance, the supposedly indivisible atoms, of which the elements are composed, were broken down into ultimate particles. Today, the atomic number of an element corresponds to the number of protons in the atomic nucleus. One consequence is that humanity has acquired a new source of power, that of 'the atom'.

RATIOS AND RATIONALITY

All this brings home the fact that science is not, in a notorious phrase, 'organized common sense'. When it depends on mathematical logic, science is very far from common sense: it depends on an understanding which comes only from taking pains with strange ideas.

In everyday, modern life, such understanding frequently concerns matters literally of life and death. During the past three centuries, mathematicians have created the systematic study of chance or probability and so have invented statistical analysis. The word 'probable' is derived from the Latin *probare*, to approve or to prove. In the Middle Ages, it meant 'sanctioned by authority' (especially of the Church). But, in the seventeenth century, it came to signify 'worthy of belief to an extent justified by the evidence': here was that rejection of authority which is typical of scientific research.

Although mathematicians began work on chance for its own sake, their findings were soon put to use: for instance, to calculate the costs of life insurance—when are certain kinds of person likely to die? and the insurance of property—how likely is a ship to sink? Also, calculations concerning rates of births, deaths and illness became essential information for the management of the state. Governments now depend on them for planning the supply of schools and hospitals and, of course, for decisions on taxes. The figures on homicide rates in chapter 3 make another instance.

Today, much statistical analysis arises from biological variation. People differ; so do all nonhuman organisms, including those that we eat and those that cause disease. To describe all this diversity, let alone explain it, requires special methods. Like evolutionary theory and genetics, use of these methods demands a well developed critical faculty. In an admirable and disconcerting study, *The Empire of Chance*, Gerd Gigerenzer and others describe how, in much of biology and the social sciences, statistical analysis is wrongly regarded as *identical* with the scientific method. So we must now see something of both its scope and its limitations.

Some mathematical findings influence the daily lives of parents (and physicians). An anxious reader may have been told of the need to immunize small children against infections such as pertussis (whooping cough), measles, rubella, mumps and other infections. The resulting protection from serious illness has been analyzed statistically on a vast scale and is not in doubt. The likelihood of a bad reaction (a 'side effect') is exceedingly remote. The proposal to immunize is therefore based on numerical findings. But it also requires nonstatistical decisions. Fortunately, these are easy to make: the prediction of a safe protection against serious illness is so reliable that few parents oppose immunization for their children.

Other predictions, also of vast importance, are of a different sort. In the 1930s, in certain countries, especially France, an outcry developed about an impending *decline* in the population. (Today, at a time of struggles to curb population growth, this notion is likely to cause mirth.) Demographers were believed to have made elaborate calculations, based on trends in birth and death rates, and to have predicted a calamitous loss of fertility and hence of total numbers. A book appeared in England with the dramatic title, *The Twilight of Parenthood*.

The calculations were indeed elaborate; but in fact the statisticians worked out not what *would* happen but what would happen *if* certain trends persisted. As they knew, both fertility and mortality can change rapidly and unpredictably. The relevant influences range from the economics of peasant farming to religious edicts about contraception.

Similar logic is needed for statements about global warming and the future of the world's forests, grasslands, waters and deserts. In the 1980s, computerized studies of climatic change revealed a likely disaster owing to the increase in carbon dioxide and other 'greenhouse gases' in the atmosphere. Yet in 1999, a newspaper had a prominent headline, WARMING IS NOT A CERTAINTY: EXPERT. But it does not require an 'expert' to tell us this. Statements on populations and global warming are not predictions of certainties: they may be overstatements *or* understatements. Nonetheless, such calculations are essential, for they make possible rational plans, based on probabilities, for controlling fertility and for interrupting the destruction of the biosphere. The survival of civilization may depend on the success of such plans.

TIME AND CHANCE

Unfortunately, as in the last examples, statistical analysis often goes not with science but with misunderstanding. The reader has probably already become familiar with other sources of confusion, such as those concerning correlations. Two things vary together—are correlated: therefore, one seems to cause the other. But, of course, it does not follow. In 1978, an American psychologist, Greg Kimble, described how somebody once investigated the numbers of mules in the States of the USA and also the numbers of PhDs. A strong negative correlation was found: the more mules to every ten thousand persons, the fewer PhDs. Does that mean that mules *prevent* people from getting doctorates? For that matter, a positive correlation exists

(I suspect) between my age, in years, and the world incidence, year by year, of assassinations and other calamities, yet I accept no responsibility for any of them.

More serious, in some multicultural communities Catholics have the highest crime rate. But that does not prove that Catholicism *causes* crime: such figures should lead to inquiry; for instance, does a true causal relationship exist (as indeed it does in the USA and elsewhere) between *poverty* and crime?

The last case illustrates two simple principles: that we must not accept correlation as identical with causation; but, equally important, we must not brush off correlations because they can be misleading. They are crucially important in human affairs. Like other statistics, even when they prove nothing they may suggest fruitful hypotheses.

An example is the horror story of cigarette smoking. Early in the twentieth century, lung cancer (primary carcinoma of the lung) was very rare. By mid century, it had become alarmingly common. The rise was precisely correlated with an increase in the smoking of cigarettes. (The relationship is, however, not like that of an infection which leads almost at once to illness: smoking was followed by cancer only after many years.)

The first well founded warning about smoking and health was published in 1950. In 1962 and 1964, the British Royal College of Surgeons and the United States Surgeon-General published reports based on meticulous, large scale studies: both stated firmly that smoking causes cancer. (Other ill effects of smoking, which also emerged, include a substantial increase in heart disease and, worse, in emphysema—a lung condition which can cause very slow death from suffocation.)

The warnings could not be derived from experiments, because these would require comparing experimental groups, subjected to cigarette smoke, with carefully matched control groups given a placebo (as on page 18). They were inevitably based on correlations. So an eminent statistician criticized them. The correlations, he said, failed to demonstrate the supposed effects: those who chose to smoke were perhaps the very people who were already liable to develop cancer and other ills. His comment could be justified as an exercise in formal logic; but it was obviously fantastic, for it disregarded the total picture. This included the nonstatistical fact that smokers breathe in a mixture of many substances. It was indeed thoroughly irresponsible (and, of course, welcomed and publicized by the tobacco companies).

Fortunately, the story does not end here. When a correlation is observed, it may lead to hypotheses which can be rigorously tested. One test arose from the decision, by many people, especially physicians, to stop smoking. Over years, the fates of those who stopped were compared with those who did not. Stopping, as predicted, went with a decline in cancer and other ills. The obverse of this finding is that illness is positively correlated with passively breathing the smoke of others.

An additional hypothesis was that study of the contents of tobacco smoke would reveal the presence of cancer-causing substances. And so it did. (For more, the reader should consult *Smoke Ring* by Peter Taylor and *The Cigarette Papers* by S.A. Glantz and others.)

Hence a mass of research has been generated by the initial findings (and has led to successful action against smoking and against the tobacco companies). The story therefore illustrates both logical principles and the difficulties met in applying them. It also shows how statistical studies can lead to a successful search for causes. In this case the outcome has been the prevention of illness and death. Other such studies range from the causes and prevention of violent crime (pages 54–56) to the methods of promoting educational success.

AN AVERAGE PROBLEM

When analysis of probabilities began, it led to the difficult concept of the average. This can refer to anything that one can measure or count, from the level of haemoglobin in a red blood cell to the number of hours a week spent watching soap operas. One outcome was the notion of the average man (or person); he was followed by that hypothetical legal being (who is always male), the Reasonable Man. According to Gerd Gigerenzer, in the eighteenth century statisticians assumed that at least a minority of men were reasonable; but, later, 'psychologists discovered to their dismay that almost no one was reasonable'. People and their social institutions, as shown on the next page, are inconveniently variable.

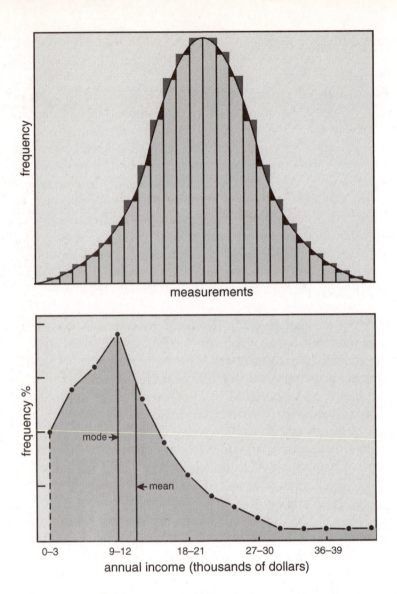

frequency

measurements

frequency %

mode→

←mean

0–3   9–12   18–21   27–30   36–39

annual income (thousands of dollars)

Above: The Gaussian normal curve of error. The curve represents the result of making many careful measurements of the same object or phenomenon: the measurements never give exactly the same result. A similar curve results from measures of different individuals in a population. *Below*: A skewed variety of the curve, shown by the distribution of incomes. The modal or most typical income does not, as it does in the top figure, coincide with the mean ('average'). The latter is dragged up by the small numbers of high incomes. An example of the limitations of averages as sources of useful information.

Today, confusion still reigns. Not even people in high places are well informed about averages. During the second world war, Dwight Eisenhower (1890–1969) commanded vast armies; later, he became president of the United States. While president, he expressed alarm, because *half* of all United States citizens have an intelligence score *below the average*. He was referring to the intelligence quotient, or IQ. And it could not be otherwise. The distribution of scores is symmetrical about the middle value or median. Look at the upper figure opposite. The method of scoring produces the result that worried Eisenhower: it allows no other outcome.

Of course, as the lower figure shows, not all averages are in the middle: some important distributions are very skewed, for instance, those of income and cricket batting scores. We know what those distributions tell us: that the likelihood of the reader being a millionaire, or of scoring a century, is rather small.

RURITANIANS VERSUS KUKUANAS

In human affairs, it is often necessary to *compare* averages and the variation around them: measurements may be of groups or populations—poor and rich; women and men; blacks and whites, fat and thin . . . Such comparisons often arouse strong emotions. To illustrate further the principles involved, I take an imaginary example.

Consider two countries, Kukuanaland and Ruritania (names borrowed from famous works of fiction). The Kukuanas vary in appearance but are said to be 'black', the Ruritanians, nearly 'white'. The character measured (highly valued in both communities) is called 'zing'. Possible scores are from 0 to 100. Zing has been intensively studied (and debated) on large numbers in both populations; but it has not been found to be correlated with skin pigmentation or other anatomical features in either population.

The distribution of zing in each community is roughly 'normal', like that shown in the upper figure opposite. The estimated mean value of zing for all Kukuanas is 61, that of Ruritanians, 57. The difference is very highly 'significant' in a statistical sense. Hence some Kukuanas are confident of their genetical superiority over Ruritanians. But this belief is disputed. First, the environments of the two populations differ: Kukuanas live in fertile uplands of central Africa, Ruritanians in a mountainous region of central Europe. Second and more important, the two cultures, including their schools, are different. Variation in zing has been shown to be much influenced by both schooling and health. Although, therefore, zing is said to have a high heritability (a technical term often misunderstood, see my book, *The Science of Life*), no valid conclusion can be drawn on the genetics of the difference between the populations.

For reliable conclusions, it would be helpful if some Kukuana and Ruritanian children were reared together, as if they were experimental animals. But few have the same schooling and even they are brought up by their own parents: the family environments of the two groups therefore differ. Nonetheless, when such children are compared, their average zing scores are about the same. Yet those who emphasize genetical differences say that these immigrants are not typical of their respective populations.

An additional complexity arises when the sexes are compared. While Kukuana men score higher than the women, the reverse is true of the Ruritanians. According to geneticists, this may reflect a genotype-environment interaction; and some Kukuana women agree: they attribute their lower mean score to discrimination against their sex, hence (they say) it reflects an environmental influence.

Most geneticists and social scientists agree that no firm conclusions are possible on the genetics of zing. More important, people interested in social action point out that, despite the 'statistically significant' differences between the populations and the sexes, the distributions have very large overlaps. Both populations and both sexes therefore have small but important minorities with very high zing (as well as some who are almost zingless). Hence concentration on statistical 'significance' can lead to disregard of common sense: for the purposes of practical action, the populations do not differ. To raise zing scores, the activists say, schooling and other relevant factors should be improved. Statistical analysis can then be usefully applied to describing the effects of such changes. The debate continues.

[The slight resemblance of zing to the intelligence quotient is of course purely coincidental.]

THE SIGNIFICANCE OF STATISTICS

This parable brings up the knotty question of 'significance'. Here is a statement, from an article by a scientist but addressed to nonspecialists, on attitudes to beauty. The author refers to responses by 52 women and 44 men, all undergraduates, who were presented with a task like that of Paris when he awarded the apple to the goddess Aphrodite: they were required to choose between faces. 'The outcome [he writes] was that observers . . . regarded symmetry in opposite sex faces as attractive, *a result that was statistically significant*' (emphasis added). Such statements now appear in the press. The reader may reasonably ask: what does the italicized phrase mean?

In fact, it signifies only that a certain type of calculation had been used to *describe* the findings. The article from which it is taken suggests that some of our aesthetic preferences are fixed by the past action of natural selection: that is, that they are 'genetic' or even 'instinctive'. But to test this we should ask several further questions, not all mathematical: just who were the subjects? what was their relevant previous experience? would a group from a different community give the same result? would a different statistical procedure give another result? None was answered in the article. Yet, without such information and more, the phrase has no useful meaning. It is an example of the danger of statistical analysis becoming a mindless ritual.

Averages offer additional problems. The bell shaped curve on page 152 is the basis of much statistical analysis. First named by K.F. Gauss (1777–1855), it is now called the normal curve *of error*. 'Gaussian' distributions are found whenever measurements of the same object or phenomenon are repeated. The figures are never exactly the same: small errors (at least) always occur. If a physiologist gives a figure (say, 37^0C) for the normal deep body temperature of a single healthy person, that figure—if it is to be reliable—should be an average derived from many observations.

A similar curve results from measures of different individuals. A textbook may give average or typical figures for the amount of haemoglobin, the number of white cells and the amounts of various proteins in a given volume of human blood. All vary, some greatly, and it is important to identify the sources of the variation. These include imperfections in the equipment used and errors by the people using it. But a departure from the average is often an early indication of illness. Physicians who use such figures are, in effect, comparing two groups—sick and healthy. When populations are compared statistically, the averages (strictly, the arithmetic means) and the scatter around the means may therefore be indicated by curves such as those in the figure. *All these are descriptions.* The calculations, sometimes very elaborate, are indispensable for reporting many experimental findings. They may also reveal relationships not at first evident. But alone they prove nothing.

Suppose that calculations lead to the statement that two populations are significantly different at the 5 per cent level. This is still a description: it is one way of presenting the different forms of two distribution curves. What action is then taken often depends on personal judgement. If the reader, wishing to perform some feat, is told—on statistical grounds—that there is one chance in twenty (5 per cent) of dying in the attempt, he or she may decide to cancel the performance. But if the possible penalty is only a small fine, the risk may be cheerfully taken. Similarly, if such a 'significant' difference is found in experimental material, a scientist may decide to perform many arduous experiments to test a hypothesis derived from the finding.

I take the 5 per cent level as an example, because it has been customary to call differences at that level 'significant'. This usage is highly misleading: a definition of 'significant' in its more usual sense is 'full of useful or important meaning'. The different uses have led to a number of widespread but erroneous beliefs. The level of statistical significance does *not* tell us:

- the size of the difference or other effect described;
- the probability that the hypothesized effect is true or false;
- the probability that no effect is present;
- the degree of confidence that a finding can be repeated.

Today journalists often write that numerical findings are 'significant', when in fact this means only that they have been analyzed statistically. Whether they are significant in its primary sense can be decided only by intelligent, critical scrutiny. To quote Gigerenzer again, 'no amount of mathematical legerdemain can transform uncertainy into certainty'.

HOW SHOULD CHILDREN BE TAUGHT TO READ?

Here is a simple example of quantitative, experimental findings. The primary objective was to compare two methods of teaching children to read. In the 'phonic' method the children speak the words they are reading; in the 'sight' method, they try to get a visual impression of each word. Children, however, vary. Two groups were therefore studied, 'gifted' and 'nongifted', assessed by tests of 'general intelligence'. The figures in the table are of average reading speeds in words per minute.

	Method of teaching	
	phonic	sight
gifted children	60	50
nongifted	30	10

Clearly, these American children did better with the phonic system. But there was also an interaction between ability and teaching method: the beneficial effect of the phonic method was much less for the gifted than for the nongifted subjects.

This 'two-by-two' experimental design simultaneously records the effects of two variables: 'giftedness' (defined by a specified procedure of assessment) and teaching method. Some researches, however, require attention to large numbers of variables. An example is the study of anger described on pages 54–56, in which the variables included many kinds of action. Even to describe them in a comprehensible form, such studies usually require elaborate statistical procedures. But interpretation of the calculated results still requires nonstatistical judgement.

MATHEMATICS: HUMAN OR INHUMAN?

Since mathematics is a product of human activity, it has a history. Reuben Hersh, who knows mathematics and also writes effectively about it, describes mathematics as an outcome not only of disinterested thought but also of cooperation and competition, of friendship, rivalry and criticism. Mathematicians discover theorems. Nobody knows how. But they do not begin with axioms. Afterwards, the theorems are shown to require axioms on which to stand, in accordance with the rules of the game.

Hersh quotes the description by Augustin-Louis Cauchy (1789–1857), a French mathematician, of how he discovered an important theorem, during a night when he was sitting up because of asthma.

> . . . at about 3.30 there suddenly arose before me the Central Theorem . . . The next afternoon, in the mail-coach I thought through what I had found, in all its details. Then I knew I had a great theorem . . . the proof was in fact very difficult. I never doubted that the method of proof was correct, but everywhere I ran into gaps in my knowledge of function theory or in function theory itself. I could only postulate the resolution of these difficulties, which were in fact completely resolved only 30 years later.

The confessions of another, still more famous French mathematician, J.-H. Poincaré (1854–1912), are quoted by René Taton in his *Reason and Chance in Scientific Discovery*. Poincaré described himself as incapable of doing simple arithmetic without error; but, like Cauchy, he also wrote of a mathematical intuition which enables a few people to guess at 'hidden harmonies and relations'. He reports a struggle, undertaken—he says—when he was very ignorant, to prove a proposition in higher mathematics. One night he drank black coffee and could not sleep; ideas 'jostled' in his head until, next day, on a journey, the crucial concept came to him (it involved nonEuclidean geometry). 'I felt absolute certainty at once; [later] I verified the result at my leisure to satisfy my conscience.'

A mathematician such as Cauchy or Poincaré can choose what problems to study but not what the answer will be. In this, mathematics differs from music and other products of human creativity. Mozart in his time was considered a musical revolutionary: when one of his masterpieces was first performed, he was told that it contained too many notes. After an early performance of Beethoven's 7th Symphony, another composer said that Beethoven was fit for the madhouse. Some of Chopin's contemporaries found his compositions exotic and incomprehensible. But none of these critics said that the music could be corrected by new calculations. In mathematics, in contrast, once the rules are settled (by human decisions), they usually have to be accepted: hence there are correct and and incorrect answers.

Also strange, unlike languages such as English and Malayalam, mathematical statements are common to all humanity. And, although they are not about anything, they can be applied to define and even to solve problems anywhere in the material world. When so applied, they present human reason in a particularly clear form. But science, though it too exemplifies human reason, is much more than mathematics. What is it?

CHAPTER 11

FIRE FROM HEAVEN

Wealth heaped on wealth, nor truth nor safety
                                         buys,
The dangers gather as the treasures rise.

SAMUEL JOHNSON

THE GREEKS TOLD HOW the Titan, Prometheus, whose name means 'forethinker', stole fire from the gods and so enabled humanity to alter the order of nature and to create the natural and medical sciences. *Homo sapiens*, as a result, is a firemaking species—the only one: *Homo incendiarius* (chapter 6). Yet Prometheus was condemned as a wrongdoer. In modern times he has even been likened to Satan, the hero of Milton's *Paradise Lost*. In the best known form of the legend, his punishment for defying the supreme god, Zeus, was to be chained to a rock, where a vulture visited him daily and gorged on his liver.

SCIENTISTS IN CHAINS

In our day, science offers steeply increasing knowledge; but much is directed not to giving humanity power over nature but to giving a few human beings power over others: many modern scientists are chained to a rock by the mindless greed and lust for power of their paymasters. Those who work in secret for the massively profitable armaments industries are the most frightening. In the 1990s one-third of all scientists in the world were employed on work related to war. Their principal gifts to humanity are the means to burn cities, to poison forests and to

160

Prometheus, the rebel: for his temerity in stealing fire on behalf of humanity, he is chained to a rock while a vulture consumes his liver.

kill the people in them. Others are employed by tobacco companies (pages 150–1).

A leading example, of scientific innovation as private property, controlled by market forces, is the grandiose Human Genome Organisation (HUGO), of which the objective is to map all the genes (said to be 60 000) on the 23 human chromosomes. According to one enthusiast, Leroy Hood, 'the information generated by the human genome project . . . will ensure the United States a highly competitive position in the worldwide biotechnology industry'. (See *The Code of Codes*, edited by Kevles & Hood.) Such a statement does not tell us how HUGO will benefit *humanity*. Worse: in the USA, patents are being taken out in newly discovered organisms, and in plants and animals into which genes have been inserted: the primary objective is private profit.

A French social scientist, J.–J. Salomon, in an essay on science as a commodity, writes of science as losing its ancient role as 'a learned and innocent activity': the search for truth, he holds, is not the goal of modern science. For him, as for others, the explosion of nuclear bombs in 1945 was a turning point. Salomon brutally likens the scientists dependent on the military or private industry to the 'Young Lady of Kent',

> Who said that she knew what it meant,
>> When men asked her to dine,
>> Gave her cocktails and wine:
> She knew what it meant . . . but she went.

An English sociologist, Steve Fuller, has similarly written about science as an institution in modern society. We think of the universities as sites of independent scientific and other research; but today, he says, we face, 'the privatization of innovation as intellectual property' in the possession of entrepreneurs and industry. Meanwhile, teaching and examining remain with the universities, which become factories for producing degrees.

A principal function of the universities should be to educate undergraduates and graduates in methods of exploring the natural world—a task of formidable difficulty, which is rarely acknowledged outside the groves of academe. To attempt this, teaching and research in the natural sciences must be inextricably combined. If they are separated, then, as Fuller suggests, 'Science' could come to be what much religion has become, 'a set of institutions and rituals'.

In the legend, Prometheus was eventually rescued by the Hero, Herakles. Modern scientists who wish to work freely both for knowledge of nature and also—in Bacon's words—for the relief of man's estate, cannot hope for rescue by a new Hero. If they are to escape, it must be by their own efforts and by those of wellwishers.

And their potential wellwishers are entitled to demand a response to the central question which I try to answer in this book: just what is this science that they practise?

PHILOSOPHICAL FACT AND FANTASY

The sciences which appear in the preceding chapters have no obvious unity. Correspondingly, few people call themselves scientists: they say that they are (for instance) astronomers or zoologists. One source of disunity is the difference between the physical sciences and biology. Although stars and rocks change with time, the laws searched for and found in the physical sciences are permanent regularities. In biology, time and diversity produce incessant, often rapid change. The perhaps thirteen million existing species of organisms on earth are the outcome of natural selection and other sources of variation. Even the individual members of a species differ from each other. Merely describing and classifying them is a monstrous task. As a result physicists, ignorant of the organic world, have been known to dismiss biology as mere stamp collecting. A historian of science, David Knight, has described how physicists, at one time, regarded even chemists as 'little better than cooks'. The obverse was seen when a prominent biologist referred to his colleagues (and himself?) as suffering from physics envy.

Such badinage may be illuminating about the frailties of scientists, but it tells us little about science. Can we find anything that binds the sciences (and scientists) together? I answer after half a century of research and of trying to teach not only zoology but also scientific principles. Inevitably my reply reflects personal experience and bias.

A *scientific method*, common to all the natural sciences, has been supposed to exist. One notion, long held by some philosophers, is of an investigator beginning with facts collected in a totally neutral fashion. Conclusions are reached or thought of only when many observations have been made. A name given to this entirely imaginary procedure is induction. A chemist, P.W. Atkins, has taken an extreme position. During a forceful sermon against religion, he states that science, to be honest, 'must start from nothing at all'. To which King Lear provides the appropriate response: 'Nothing will come of nothing; speak again.'

A contrasting scenario has been presented by K.R. Popper, based on what he believes scientists actually do. It is an agreeable picture. They first make 'marvellously imaginative, bold conjectures'; these, it seems, come from nobody knows where.

They or their colleagues then test them by observation, experiment or argument and, in doing so, they try to refute them. Popper holds that statements should be called scientific only if they are falsifiable. His great work, *The Logic of Scientific Discovery*, is based largely on the history of physics. His picture of the growth of scientific knowledge is a logician's vision. Studied carefully, it may provoke salutary thought; but it has little relationship with the day to day work of most scientists.

CLASS STRUGGLES

The activities we call scientific include both classifying and experimenting. Classifying, we know, comes first (page 141); and for decades, indeed until recently, it dominated the biological sciences. Many biologists are still systematists (taxonomists). Although, today, they do elaborate experiments, their findings concern the existence and naming of organisms. Such statements are not falsifiable in the same way as are those about causes, yet they fall within science as it is usually understood. Moreover, whether we classify the classifiers as scientists or not, it was their systematic study of organisms that led to Darwin's revolutionary concept of natural selection.

Appropriately describing, classifying and measuring phenomena is not easy. Nor is it a prerogative of biologists. In the physical sciences, the most obvious case, not always noticed by philosophers, is the ordering of the stars by astronomers. Others are the arraying of the elements by chemists (page 147) and the classifying of rocks by geologists.

Despite the rise of experimental biology, the importance of classifying has not diminished. News items below, taken from 1998 alone, show the increasingly urgent need for extensive, accurate cataloguing of the world's organisms.

- The International Union for Conservation of Nature and Natural Resources has listed 34 000 of the known 270 000 species of plants as endangered by the food, wood and pharmaceutical industries.
- One-tenth of the 9000 species of birds and all free-living cats are threatened with extinction.
- A report by the World Wide Fund for Nature, entitled *The Living Planet*, describes the processes which have already

wiped out fisheries in great lakes in North America, Africa and Asia. These processes are being accelerated: as a result the water supplies of human populations too are disappearing.

- In the twenty-first century, malaria—already one of the world's three most serious infections—is expected to spread much more widely, carried by mutant *Plasmodium* parasites and *Anopheles* mosquitos.
- The Amazon rainforest and large parts of eastern USA and southern Europe are threatened with conversion into deserts.

Many organizations, and some governments, are working to arrest these calamities. To do so requires a massive increase in the numbers of biologists, including classifiers.

THE SEVEN HUNDRED YEASTS

Here is a sample of what is already being done. Among the millions of species of organisms are the yeasts—a tiny group of microorganisms of which, by November 1998 (so I was told), only 702 species had been described. Many more are believed to exist, in the soil, on plants and elsewhere. Today, we not only depend on them for cheese, wine, beer and risen bread: yeasts are also recognized to be biochemically versatile, easily grown in the laboratory and readily modified genetically. They are therefore much used for industrial production of valuable drugs. An example is the use of *Pichia angusta* to make hirudin, an anti-clotting agent. They can also produce antigens—for instance, against a form of hepatitis.

The names and characteristics of yeast species are provided by the exertions of systematists in many countries. The world's largest collection of yeasts is held by the Centraalbureau voor Schimmelcultures in the Netherlands. The work of the Bureau is of great economic importance and is supported by the Dutch government. It identifies the specimens it receives and provides general information about yeasts and other fungi. It is a model of scientific organization which should be imitated on a scale far greater than at present. The number of classified yeast species is something like 0.025 per cent of all known species. This figure conveys a warning on the size of the effort needed to classify them all.

(For a classification of all organisms, see *Five Kingdoms*, by Margulis and Schwartz. On yeasts, see *Yeasts: Characteristics and Identification*, by James A. Barnett and others.)

EXPERIMENTING

The physical and biological sciences have classifying in common; but, more obviously, a practice central in both is experimenting. Experimental projects begin, not with collections of facts, but with problems. Often, these are imposed by government or industry. If, however, a scientist is free to choose, a problem may be suggested by something observed, especially if it is surprising. For example, very small mammals have colonised apparently lethal environments at $-10°$ Celsius. How did they do it? (Appendix 1 gives a partial answer.)

It is easy to distinguish two kinds of experiment.

- When I grew geraniums in window boxes, caterpillars sometimes infested them and ate the leaves. The larvae seemed to be spread out. What would happen if several were put on the same leaf? Until recently, 'experiment' meant this sort of exercise of curiosity: today a critic might ask, what would be the point of doing it?

- The alternative is an experiment in the modern sense. It begins with a clearly stated hypothesis—for instance, that if two caterpillars are put on the same leaf, each will secrete a pheromone causing the other to retreat. Such a proposal could arise from knowledge of species which display intolerant behaviour. The outcome might help to explain the activities of these larvae—and, perhaps, also those of other species. (It might even be used to protect the plants from the pests.)

It is sometimes suggested (though seldom, I think, by people who do experiments) that only these two kinds of experiment exist and that only the second is worthy of respect. It is indeed satisfactory if one can propose a rigorously stated, testable hypothesis; and better still if one can test it; and, best of all, if the findings match what has been predicted. But research is hardly ever so tidy. As Francis Crick has written, 'In research the front line is almost always in a fog'.

Joseph Priestley (1733–1804), who—like Crick—made historic advances, was a maverick Unitarian minister famous for his experiments on gases. He designed new apparatus and with it discovered what we now call oxygen. He also demonstrated the dependence of animals on oxygen and the release of oxygen from plants. Charles Singer gives a lively account of how he emphasized the role of chance in discovery. In his experiments, he said,

> I was so far from having formed any hypothesis that led to the discoveries I made . . . that they would have appeared very improbable to me had I been told of them; and when the decisive facts did at length obtrude themselves upon my notice, it was very slowly, and with great hesitation, that I yielded to the evidence of my senses.

Many lesser scientists must have had similar experiences. The trouble is, there is *no* logic of scientific *discovery*. Despite the title of his famous work, Popper acknowledges this. He distinguishes, he says, sharply between the conception of a new idea (which has 'an irrational element') and examining it logically. A new idea can emerge either before experiments begin or, more likely, while they are in progress.

A philosopher, W.H. Newton-Smith, in a discussion of scientific method, writes that the good experimenter, 'sadly ignored by philosophers of science', is always making decisions on what to do next. This skill is not described, let alone explained, by any theory or by logic. It is intuitive (page 139). Since discovery is an outcome of human intelligence, we should not be surprised at our ignorance about it. Despite the exertions of psychologists, we have no general scientific laws concerning human action. (I say more on this in *Biology and Freedom*.)

Priestley's work also reminds us that experimenting usually demands apparatus, sometimes complicated and, today, often expensive. Moreover, experimenters commonly have the experience of modifying or even designing equipment. When I wanted to study the exploratory behaviour of small mammals over long periods, it was disconcerting to find that no suitable equipment existed. So I devised a computerized 'residential maze' which, later, was used also by others. Its design arose largely from nonlogical thinking combined with trial and error. But the experiments themselves, and the analysis of the findings, were logical. This sort of thing is an everyday happening in laboratories. Its importance is recognized by historians of science, rarely by philosophers.

One aspect of experimental logic we already know: the control (chapter 1). Experimenters look for causes by studying differences. Typically, the effects of two conditions are compared: in the example of a drug believed to cause sleep, the only difference of the experimental from the control group should be taking the drug (which is difficult to arrange). A more complex instance concerns teaching children to read (page 157).

All experimental researches demand that, once the problem has been identified (clearly or not), many particular events are meticulously recorded. As every struggling experimental biologist knows, up to a point the greater the number of observations, the more likely will the experiments lead to a convincing conclusion. This may be regarded as a truth contained in the idea of induction.

Pride and Prejudice

To understand discovery, even imperfectly, we must also acknowledge what motivates it. Especially when they are not under orders, scientists may be influenced by fascination with phenomena such as flowers or faces or mountains or stars. A microscopist may find beauty in tissues and cells and feel pride in the skill used in revealing it.

Private passions can therefore lead to systematic study and scientific findings. The Spanish physician, Santiago Ramón y Cajal (1852–1934), was an outstanding histologist and a founder of modern neuroanatomy. In an enjoyable autobiography, he describes his 'honeymoon with the microscope' during which, untaught and even discouraged, he did nothing but satisfy his curiosity. 'There was presented to me a marvelous field for exploration, full of the most delightful surprises' [such as the cells on page 26]. Later he came to realise the need to discipline his 'vagrant curiosity' and to fit it to the rigid frame of 'a programme drawn up beforehand'.

Whenever a scientist (or anyone else) perceives objects or events, that perception is biased by attraction or expectation. As soon as we observe something, we impose on it our own patterns derived from many sources. When I began teaching zoology, I was dismayed by the inability of students—even the few who could draw quite well—to represent what was before them.

We impose interpretations on everything we observe. 'Mach's corner' consists of lines and shaded areas on a flat surface but it is seen as a solid object. In this case interpretation is not constant. Look at the figure steadily for a few seconds.

Often they produced something like a textbook diagram. 'Draw only what you see!', I exclaimed. Later, it sank in that I was demanding almost the impossible, like the teacher of art portrayed by E.H. Gombrich.

> The academic teacher bent on accuracy of representation found, as he still will find, that his pupils' difficulties were due not only to an inability to copy nature but also to an inability to see it.

For scientists' accounts of this phenomenon, consult Richard Gregory's *The Intelligent Eye* and M.L.J. Abercrombie's *The Anatomy of Judgment.* All we perceive, even if it is quite simple, we interpret. When we are shown the Mach corner above, what is before us is a flat page with a few lines and some shading; but what most people *see* is a figure in three dimensions. And, if it is steadily observed, the figure changes. In this case, our interpretation fluctuates. All paintings and drawings (page 123) make use of our compulsion to interpret what we observe.

Similarly, when a person is faced with a problem, what is perceived is influenced by private, nonlogical factors. To return to my geraniums: let us suppose I had asked a sample of students what determined the caterpillars' choice of leaves and flower buds, what kept them spaced out and how they interacted. Among possible responses, one would be to record social responses—those affected by other caterpillars. Another, influenced by the many studies of insect orientations, would treat the insects as mechanisms with sensors and neural connexions ('hard wired'!) which force them to respond to the direction of the light or to the odours around them. A third would be to calculate the flow of substances and energy through them—a procedure of great ecological importance in other environments.

All experimental projects are likely to be influenced in such ways, especially by knowledge of the work of others. What they also have in common is meticulous observation, accurate reporting and comment (sometimes logical) by colleagues.

PREDICTION OR EXPLANATION?

Experimental findings are commonly said to make prediction possible. Sometimes they do. And this is very important. But much scientific research, experimental or not, is better described as seeking satisfactory *explanations* of puzzling phenomena. Here are three leading examples.

- First is Newton's theory of gravitation. Today we are accustomed to this revolutionary concept: we take it for granted that the force which causes apples to fall off trees is responsible also for the tides and for keeping the moon in its orbit. We learn too that Newton's *Philosophiae Naturalis Principia Mathematica* (1687)—sometimes described as the greatest of all scientific works—made possible prediction of the movements of the stars in their courses.

    Yet an algorithm for calculating, say, the orbits of the planets, can be devised (and, today, stored in a computer). It then allows accurate prediction of planetary movements but it does not explain them. Newton's great work *explains* what we observe in the heavens and elsewhere: it gives us understanding. Predictability and understanding are not the same.

- Another momentous concept, natural selection (chapter 6), still causes confusion. It is commonly referred to as a theory, yet it is quite unlike the theories and hypotheses of physical science. Darwin *uncovered* a process in nature: despite the inaccessibility of past events, he and Wallace showed how genetically determined variation can result in evolutionary change. The idea of natural selection provides a framework for the whole of biology: it explains much; and it encourages us to ask (and, occasionally, to answer) what are the functions of particular features of organisms. But not even the latest forms of Darwinism give us detailed predictions of anything.

- The third we owe to a superb experimenter. The French chemist, Louis Pasteur (1822–1895), finally established the principle that life is not generated from nonliving matter. This major demonstration of a regularity in the living world is, like natural selection, a finding of immense explanatory power. (It also helps us to predict results of infection with microorganisms.)

These founding principles are each linked with innovators whose work an ordinary scientist may study with awe. Yet their conjectures were not simply the products of individual thinkers, 'voyaging through strange seas of thought alone'. Like the rest of us, Newton, Darwin and Pasteur built on what had already been constructed. Newton himself said that, if he had seen further than most men, it was because he had stood on the shoulders of giants.

The daily work of a modern experimentalist is not only founded on past knowledge. Like that of the imagined botanist in chapter 8, it is nearly always provoked by small problems and carried out with others. And, again like that of the botanist, it may require both reduction (for instance, to genes or to chemistry) and also attention to wholes. If the conclusions convince colleagues, they appear as reports in scientific journals—tiny fragments in the great mosaic of scientific knowledge. In that mosaic, they are always open to be altered or replaced. But, even if they are eventually discarded, they have helped to advance knowledge.

When experimenters look for causes, they assume some degree of uniformity in nature: they expect order in all the apparent disorder around them and, with good fortune, they find it. Some, therefore, are tempted to assume that all events are determined in advance. But, as we know from chapter 5 (and from modern physics), we are not obliged to accept the idea of a fixed destiny—imposed, perhaps, at the time of the astronomers' Big Bang rather a long time ago. Our predictions are, in some fields, steadily improving; but they do not demand determinism.

This is crucially important, for reliance on extreme determinism encourages not only resignation but inertia. If the outcome of our actions is unavoidable, regardless of our wishes, we may feel that we are relieved of responsibility. Determinism, like astrology, can be an excuse for doing nothing. A reader who wishes thus to drift on the tide of circumstance should have stopped reading this book many pages back; for in the background of scientific practice, especially experimenting, is the presumption that we are not puppets of impersonal forces. Knowledge gives us power to control what happens.

The existence of this power does not, however, guarantee satisfaction with science; nor does it ensure that those who are satisfied with science have a just estimate of its scope. Here are some relevant questions.

QUESTIONS

Are critics justified in being disgruntled with science?

Earlier I mention real evils arising from technical achievements. The most notorious, still looming over humanity, arise from atomic power. Spencer Weart writes of millenarian hope, contrasted with 'apocalyptic fear, and anxiety about the revelations of knowledge' after the destruction of Hiroshima and Nagasaki by nuclear weapons.

Another source of hostility to science is disappointment. A showy research program can result in fiasco. In *Prometheus Bound*, a physicist, John Ziman, suggests that knowledge of expensive failures can cause hostility to science. For many years the United States government spent hundreds of millions of dollars on the 'war on cancer'. Meanwhile, the incidence of most cancers has continued to rise. Other failures, so far, include the hunt for a vaccine for malaria, the attempt to design translating machines and the search for power from nuclear fusion.

But such failures should not influence people against science: they should instead remind us of the unpredictable nature of discovery. This upsets politicians and managers, for they commonly demand projects with rapid and predictable outcomes. Such an attitude would have led the backers of the explorer, James Cook (1728–1779), to demand an accurate map of the South Seas *in advance* of his voyages.

Among the unplanned technical achievements of modern times are the internal combustion engine, the aircraft and automobile, electric light, the phonograph and the transistor. All these arose, *unpredictably*, in advance of use, as a result of curiosity or tinkering. (They also led to vast increases in material wealth.)

No doubt, necessity is sometimes the mother of invention; but, equally, as George Basalla points out, inventions can *create* necessity. During much of the nineteenth century, people were without any of the items listed. Today, where would we be without electricity? Many readers would, I suppose, regard it as essential for a normal existence. (And there is more to come: as I write, we have the encroachments of mobile telephones . . .)

If progress in knowledge is desired, scientists must be able to make independent findings. These must be put before colleagues for criticism—and, of course, praise. Exposure to criticism demands free communication in publications and meetings. Hence arises the need for an unfettered community of scholars—a need often much resented by politicians who manage states.

The resulting benefits and hazards have then to be scrutinized separately in the corridors of power. There the decisions taken are *political* in the sense of *related to the management of the state*. In an authentic democracy, the decisions would reflect the needs and desires of the citizens. No such utopia has, however, yet been achieved.

Alternatively, are its supporters in danger of overestimating science?

The power conferred by science, instead of arousing fear, can lead to selling science as a kind of knowledge which overrides all others. It is easy to put 'too high a value on science in comparison with other branches of learning or culture'.

The phrase quoted is from Tom Sorell's *Scientism*, which discusses what he calls 'the infatuation with science'. When findings about animal life are applied to human society (biological naturalism), the danger of overrating science is obvious. To find the truth about the human condition, we are sometimes told, we need only to reduce humanity to biology. In chapter 6 and elsewhere, I try to avoid this error by showing, with examples, how history and social science are kinds of knowledge, outside the natural sciences, which are needed if we are to understand ourselves.

A fundamental objection to naturalism appears if we examine moral conduct. Morality is part of culture. It changes rapidly, without waiting for genetical variation. Change occurs partly because moral principles are debated: they derive their impact from the customs of a society; and custom is sometimes an outcome of reason.

To support these generalities, here are instances of recent or current moral controversy and progress. Each is related to human death and survival; none is explained by biology or by any other branch of science.

- In the distant and even the recent past, owning persons, that is, slavery, was accepted without argument. In some communities, slaves could be beaten, raped or killed at the whim of their masters. Today, in most countries, slavery is illegal and regarded as horrible. To achieve this, many reformers urged freedom for slaves on moral grounds. In doing so, some lost their lives.
- Similarly, the total subjection of women to their husbands, once regarded as part of human nature but now abolished in many countries, has been and is opposed as morally wrong. This opposition, which has often provoked violence, cannot be explained by a drive to improve the reformers' Darwinian fitness.

- In the 1990s, in some countries, increasing numbers of religious leaders have opposed the custom of equating the Good with the Profitable. The reader may disagree with them; but neither agreement nor disagreement should be based on assertions of a human nature imposed by our biology. Present attitudes to private property (pages 48–49), and to the accumulation of wealth, are modern phenomena. Like the previous two instances, they cannot be described, let alone understood, except in the context of social history.

Should we assume the existence of objective knowledge?

Scientific statements assume the existence of a world which is independent of human desires and action. The reader may find this assertion no more than common sense, yet it is sometimes rejected.

- One form of rejection is the idea of women's science versus men's, black science versus white (or yellow), and—perhaps—fat science versus thin: a notion which seems to imply that the truth of scientific statements is influenced by who is uttering them. Suppose that a parent visits a school, wishing to thank the science teacher for inspiring teaching; and suppose that the teacher proves to be an extraterrestrial hermaphrodite with purple skin and four eyes. This would not invalidate gravitation, Boyle's law, Ohm's law, the conservation of mass-energy or the other principles which had been so skilfully imparted.

    The idea of women's science shows the importance of distinguishing *selection* of problems from scientific *findings*. Above I describe how individual bias can influence the choice of research problems or methods. Recently, much notable research on monkeys and apes has been by women. Some has been on aspects of behaviour, especially of females, which had formerly been neglected (by men). But the researches are attempts to find out what is true about the behaviour: their conclusions are judged in exactly the same way as are all other findings, regardless of the wishes, sex or other traits of the authors.

- In a related confusion, some people hold that, if scientific findings or theories are impelled by or used for wrong ends, then the science itself is invalid. This, of course, is nonsense. The *correctness* of scientific conclusions must not be assessed either by how they are used or by their contribution to human survival—or death. The criterion of correctness is whether the statements match the events observed.

- Nor should scientific findings be judged by the social conditions which produced them. The standard example is natural selection. Darwin's writings show him to have been influenced by prevailing ideas about competition and violence in human society, reflected in Tennyson's notorious phrase, 'Nature, red in tooth and claw'. Darwin, we know, used many metaphors; even 'selection' is derived from stockbreeding (compare chapter 9). None of this invalidates the concept of natural selection.

A philosopher, Anthony O'Hear, has discussed the writings which seem to state that scientific theories are determined, not by the realities of an external world, but by the outlook and attitudes of their own time. He calls them 'naturalistic and denigratory analyses' of science. To be consistent, the writers should apply their method also to themselves, but they rarely do so. If they did, their arguments too would have to be regarded as no more than byproducts of the 'spirit of the age'; they would then have no objective validity. Such writers are therefore in the same boat as sociobiologists and ultra-darwinians who seem to explain all human action, including their own speculations, as predetermined by genes or natural selection (pages 98–100). It is hardly surprising that the writers I am criticizing often recoil from the implications of what they write, for such a rejection of reason leaves the findings of science, including those of the writers themselves, without rational justification.

CONTRADICTIONS?

The presumption of an objective world of discoverable objects and events should go with recognizing ignorance. In 1998, a newspaper announced AN AMBITIOUS THEORY OF EVERYTHING. Below the headline was a picture of Darwin and an article on current writings. Whatever was intended, the headline implied an absurdity. Science leaves no scope for boasting that we have all the answers. Not even the word 'science' can be formally defined to everyone's satisfaction. When we try to give a balanced account of the activities we call scientific, we find a set of apparent contrasts.

- Science consists of a search for truths. Yet the more we attempt, the more we must acknowledge that finality is unattainable and that scientific conclusions are often statements only of what is probable.
- Scientific scrutiny is rigorously logical and, whenever possible, quantitative. Yet observation and discovery are nonlogical and intuitive.
- Scientists are concerned with phenomena which exist independently of human action. Yet scientific action itself occurs in a social environment and arises from moral and other social demands: scientific knowledge and methods have a history.
- Scientists are professionally required to display objectivity, whether they are studying human beings, slugs or galaxies. Yet they are themselves human beings, with the usual range of virtues and defects.
- Important discovery is most likely in a community of scholars independent of social pressures. Yet the resulting findings are always liable to be used for human ends—good or ill.
- The professional community (or culture) to which scientists belong is usually seen as set apart from other learned communities. Yet all scholarship is a search for knowledge: boundaries between disciplines—for instance between science, history and philosophy—are blurred and can be crossed.

All these statements are, I believe, valid. They bring out the lack of unity in the activities we call scientific. Moreover, the mundane applications of science are of immense importance and may reasonably seem to the reader to be its most prominent feature. Science is therefore sometimes said to be not a set of principles, or an ideology, but a useful practice, like medicine or farming.

But it can also be seen as something more.

12

CODA: THE TIMES TO COME

On a huge hill
Craggéd, and steep, Truth stands, and hee that will
Reach her, about must, and about must goe;
And what the hills suddenness resists, winne so
. . .

JOHN DONNE

FOR SCIENTISTS, DONNE'S HUGE hill is infinitely tall. Fortunately, it is better to travel hopefully than to arrive. Our final question is: in what direction should science travel?

To answer, the two constant features of science must be accepted: scientific research is a journey into the unknown; and its reported findings should match what happens in nature, regardless of our wishes. The 'cold' objectivity of scientists arises from the obligation to respect the truth about the world as it appears before us.

Yet the quest for objective knowledge is a social activity. Moreover, it matches the modern concept of democracy: the findings from research may be expected or they may be upsetting; but all are judged by a community of fellow workers, without resort to the authority of ancient scriptures or of modern High Priests. In some countries today, the need for independence for scholars is partly recognized: in them, a dissident is unlikely to be chained to a rock, burnt at the stake, beheaded or even imprisoned. (Secrecy, conformity and censorship may, however, be enforced by the demands of political bias or of monetary profit.)

Hence some scientists are still free, as they should be, to speculate, to choose the phenomena they study and to publish their work, even if the findings are unforeseen and discon-

178

certing. These privileged people can base their lives on the principles that the search for knowledge and understanding is beneficial in itself, that it is a universal feature of humanity and that it can be carried out by any kind of person.

Freedom, however, goes with social obligations. One has been stated by Carl Sagan: it is, he says, the particular task of free scientists

> to alert the public to possible dangers, especially those emanating from science or foreseeable through the use of science. Such a mission is . . . prophetic. Clearly, the warnings need to be judicious and not more flamboyant than the dangers require.

Sagan is here concerned especially with the hazards of applied physics, above all those that arise from 'atomic' bombs and nuclear power.

A second obligation is to persist in the relentless criticism which I emphasize in the early chapters. That emphasis is made necessary by the continuing prevalence of magical beliefs and, worse, hokum masquerading as science.

But, in a world where starvation and preventible disease are widespread and increasing, the overriding obligation is to try to solve our most formidable longterm problems. These are unlikely to be overcome by dramatic discoveries. The major achievements of science, especially biology, include fundamental but undramatic general ideas. One is the concept of ecology, that is, the analysis of living nature as an indescribably complex, interacting system of organisms. In the twenty-first century, the most momentous task for humanity is to arrest the destruction of the biosphere, to manage what remains and to guide its future growth. The world community is entitled to require that, in this emergency, many scientists, of all kinds, should be mobilized, as in war, to protect both humanity and the environment.

In a true democracy, the uses to which scientific findings are put would then come, if only indirectly, under popular control. To achieve this, an unprecedented increase in public understanding is needed; and managers of states, already overburdened with intractable tasks, must learn to understand the messages of science.

As a result, interpreters of science in the media carry a

massive responsibility. Nor may scientists themselves draw aside from social issues. Eric Ashby (1904–1992), a plant physiologist who became an admired university administrator, faced this difficulty. He wrote:

> One cannot consider the social function of science, without meddling in morals and politics. The present condition of mankind requires some scientists to get outside the framework of their science and to influence its interactions with society.

That is, they must themselves become interpreters.

The contributions of scientists arise not only from their special knowledge but also from their attitudes to problems. Scientists are sometimes said to be distinguished by their optimism. I do not know of a survey on whether scientists are happy about the future, but some basis for optimism does exist for those who do research or who teach science. Despite the frequent setbacks met during original studies, even those who struggle with the most exacting projects can hope to build on past achievements and to add to knowledge. They can expect to solve some of the problems they face.

Can such optimism be extended into human affairs? Today, thoughtful people have small cause to be cheerful. The real advances, moral and material, of our time are obscured by news of unremitting economic and political crises and of war and violent crime on a vast scale. Young people, who have known nothing else, are likely to accept a chaotic and irrational global society as inevitable.

But it is not inevitable. The facts of social history imply that it is still possible to improve the human condition. To hold that human society can be improved is also a moral necessity: without this, failure is certain.

Some of the ways in which biology can immediately achieve improvement I have suggested in *The Science of Life*. But biological science can provide more than piecemeal remedies. If we are to work for a better future, we need some notion of what that distant future should be. The necessary changes have begun. Worldwide movements exist, small and large, local and global, designed to save nature on behalf of humanity. So I now end with what may seem little more than a fantasy.

We are products of evolution, yet we are capable of

controlling it. To do this, we need to know much more, than we do at present, about the millions of species around us: not only their appearance but also how they survive. Learning about them and how to live with them is a task for our descendants of which it is impossible to see an end; but we can imagine a future, certainly far off, in which this task is a central feature of human society.

At the beginning of chapter 1, tribal people (formerly called primitive) appear as merged in empathy with the creatures and plants around them. For people who still live as our ancestors did until a few thousand years ago, the natural world is sacred. The survival of our civilization may depend on combining the scientific respect for truth with their respect for nature.

> Trace Science then, with Modesty thy guide;
> First strip off all her equipage of Pride,
> Deduct what is but Vanity or Dress,
> Or Learning's Luxury, or Idleness;
> Or tricks to show the stretch of human brain,
> Mere curious pleasure, or ingenious pain; . . .
> Then see how little the remaining sum,
> Which serv'd the past, and must the times to come!

APPENDICES

APPENDIX 1: THE CASE OF THE ESKIMO MICE

THE MODERN THEORY OF evolution was founded without experimental evidence. Today we have examples of adaptive change, on a small scale, in natural populations, and many experimental studies of selection. The work summarized here arose from a problem in applied zoology and is on a species of worldwide importance for public health, food storage and agriculture; but it is itself an example of an 'academic' or 'pure' study of adaptation.

✳✳✳

Domestic mice are convenient for experiments on selection. A female can become pregnant at six weeks and produce six or more young about three weeks later; a second pregnancy can begin a few hours after the birth of the first litter. Selecting a genetically mixed stock for large size can produce a substantial rise in ten generations. Such experiments mimic breeding cattle for beef on the hoof and can perhaps suggest ways of helping stockbreeders. Experiments which aim at a single, measurable feature, such as muscle weight, rarely, however, match what happens in nature. For this we need wild-type mice and experiments of a different design.

During the second world war, thriving colonies of mice (*Mus domesticus*) turned up in meat cold stores kept at about –10⁰C. They bred in the meat carcasses and used the wrappings to make nests. Kidneys were among their favourite foods. How did these small, unwelcome creatures (with an adult weight of perhaps 20 grams) adapt to such arctic conditions? They not only had to keep their own deep body temperature at about 38⁰C: more difficult, they also had to keep their exceedingly small young alive. The smaller the body, the more difficult it is to maintain a body temperature above that of the surroundings.

After the war, in Glasgow University, a foundation—endowed by a rich manufacturer of automobiles—financed experiments on house mice kept for many generations in refrigerators. At first, domestic mice were used, because it had been authoritatively stated that wild mice would not breed in captivity. Fortunately, a quiet Scottish technician, Raymond Stoddart, had not read the literature. He trapped mice on a farm and paired them off in the laboratory, where they bred—like mice.

Some of their descendants were 'selected' by being left to breed in a temperature near zero Celsius. And, after only ten generations, the resulting 'Eskimo' mice were heavier and hairier; they produced larger litters; the females secreted more, and more concentrated, milk; one kind of adipose tissue was heavier; their kidneys were larger—a correlate of enhanced metabolism; and they looked after their young better.

In such a situation, the interaction between generations is not simple. The genotype of an 'Eskimo' female influences the uterine and nest environment she offers to her young; she also grows up in a cold environment and adapts to it physiologically. Her young too have their own distinctive genotype which helps them to adapt physiologically to the cold and to respond to the maternal environment. Members of a litter also interact. Hence the features of tenth generation mice in the cold reflect their own genotype, their response to the cold and also the genotype of their mothers and even their grandmothers. The males too were very parental. The unit of selection was therefore, at least, the family group. All these interactions were analyzed.

In one paper I suggested that male house mice, treated in this way, might one day produce milk. My colleagues thought I was joking; but since then C.M. Francis and colleagues have found males of a fruit bat (*Dyacopterus spadiceus*) to be capable of lactating. Why should the males of just that one species, but not the versatile house mouse, be so odd? We can only guess; and we cannot test our answers.

All this required many experiments in which hundreds of matings were set up; it also entailed cross-fostering of newborn litters between different parents and hybridizing between different stocks. It took years of work, sometimes—as usual in research—with disappointing or incomprehensible findings.

Do mice in refrigerators tell us anything about evolution? They present several problems. Were they an example of evolutionary change on a minute scale—a change which could lead, perhaps after hundreds of generations, to the appearance of a new species? Or were they merely temporary adjustments, easily reversed? What would happen if some of the selected large, fat, fertile mice were released in a warm environment? Would their descendants replace the existing population; or would they revert to a more typical form; or would they die out? Results of some laboratory studies by others, on domestic mice, suggest that they would revert; but the only readily predictable outcome of such an experiment would be its unpopularity with the human inhabitants.

As we know, another enigma, fiercely debated by evolutionists, is the role of chance variation or 'drift'. The species *Mus domesticus* is believed to have originated in Persia. Island populations, including those in cold climates, arose from mice carried by sailing ships. Up to a point the conditions in my experiments corresponded to what must have happened to some groups during the spread of mice over the world. Occasionally, no doubt, after ancestral mice had left their sheltered quarters on board, and faced the gales and frosts of subarctic or subantarctic conditions, few remained—conceivably, only one pregnant female. At most, only a few would be founders of a population which, later, numbered many thousands.

That population would consist of survivors of rigorous 'selection' in adverse conditions; but the genes with which they began would be those of the founders. Major founder effects in my experiments were, however, made less likely by repetition of the Scottish experiments in Australia, with mostly similar results. The summary above combines findings from both hemispheres.

A third enigma concerns adaptedness. The special features of the Eskimo mice were, the reader may think, all obviously adaptive—improvements in their 'fitness'. Perhaps they were; but we are not entitled to assume it: it is often impossible to tell which of an organism's distinctive features are essential for survival and which are merely byproducts of adaptive change.

The findings on these mice do, however, illustrate one general and important feature: that, in almost any population of organisms, substantial genetical variation exists and persists. The variation can be a source of adaptive change when a new environment is encountered. The genetical variation available at one time is, however, unlikely soon to produce a new species. This requires many mutant genes (or large changes in chromosomes) and, often, a period much longer than one human life. It is possible to imagine a future world community in which enquiry into nature is so valued, that it endows researches designed to extend over many human generations.

(See the review by Barnett & Dickson published in 1989.)

APPENDIX 2: HAS DEATH SURVIVAL VALUE? A CASE HISTORY OF REDUCTION

HERE IS AN EXAMPLE, from an actual project, of the scope and limitations of reduction as a method. It has also, like the previous case, lessons concerning the time scale of research.

During a hostile social encounter, an animal falls over and dies, though unwounded. Routine examination by a pathologist reveals no cause of death. A physician faced by a patient with a baffling fever may speak of 'pyrexia of unknown origin', or PUO. I therefore called this phenomenon 'death of unknown origin', or DUO. It was first described in detail in the common 'brown' rat (*Rattus norvegicus*); it was later found in other species of the same genus; and a German zoologist, Dietrich von Holst, has studied it minutely in a Southeast Asian species, the tree shrew (*Tupaia belangeri*).

When people are asked to propose an explanation for DUO, some make 'psychological' suggestions, such as humiliation or hopelessness. An American psychologist published an elaborate comparison of DUO with 'voodoo death', in which a human being dies as a result of being cursed by a witch doctor. An obvious objection is that voodoo death, if it does occur, depends on the human capacity to respond to speech and to symbols. Rats do not go in for symbolism.

Detailed study of the rats' behaviour showed the dangers of treating the animals as if they were human (anthropomorphism). In early experiments, groups of males were put together in large cages. Nothing exciting happened. But, when similar groups were set up with females added, many males died.

So, clearly (it seemed), the males were fighting for the females. But patient observation, of small, stable groups, showed the contrary: a female in oestrus accepted any male, sometimes several; the males did not clash but took turns. In further experiments, however, female companionship was shown to enhance the readiness of a male to attack an intruder. This was evidently a form of territorial behaviour.

So far, the account is about whole animals. An obvious question was: what caused DUO? To attempt an answer, some form of reduction was needed. Was the rats' collapse due to their running out of fuel? Glycogen in the liver, which is a source of sugar, proved to be low; but sugar in the blood was, if anything, rather high. Evidently, the rats were adapting to the need for violent exertion—a phenomenon well known to physiologists. Another example of adapting was that the adrenal glands of all rats involved in prolonged exertions were enlarged. No cause of death emerged from these findings.

Then, for various reasons, we turned to the kidneys; and there at last my colleagues and I, and others who studied other genera, found clear pathological changes correlated with 'social stress'. In rats and tree shrews under attack, a mild, usually harmless kidney infection evidently flares up and may even become fatal.

Of course problems remained. As we know (chapter 6) it is often presumed that all the features of an organism are products of natural selection. If so, in this case, we are faced with the question: what is the survival value of death? Is DUO a means by which overcrowded populations are reduced? This would imply that some animals die for the good of the species ('altruistically') and so seems unlikely.

Perhaps DUO is an example of 'correlated variation' (page 85), that is, an indirect consequence of an advantageous physiological feature. A similar problem arises with other small mammals, especially voles, deer mice and lemmings, of which the populations fluctuate with some regularity. Some species become exceptionally numerous about every four years and then quickly collapse. What causes the collapse?

Research on this phenomenon was begun in the 1920s by C.S. Elton (1900–1991) and is still going on. It requires a combination of reduction with study of the whole animals in their interactions among themselves and with their environment. It is typical of the biological problems which have to be solved if we are to manage the biosphere instead of destroying it.

(See Elton's great work, *Voles, Mice and Lemmings*; and my review of DUO published in 1988.)

GLOSSARY

abiogenesis 'Spontaneous generation' or the development of living from nonliving things. No longer accepted as possible (except in the laboratory?). But living things are assumed to have arisen from inanimate matter in the remote past.

adaptability The ability to become adapted (especially to a variety of conditions).

adaptation In biology has two important meanings. 1. Physiological or ontogenetic adaptation is a change in a single individual which enables it to cope better with its environment: for instance, acquiring resistance to disease; enlargement of muscles with exercise; learning the way about. 2. Genetical or phylogenetic adaptation occurs when a population changes genetically and so increases the chances of its survival.

In addition 'an adaptation' may mean a trait which contributes to an organism's adaptedness (a usage not recommended).

adaptedness The state of being equipped to survive and breed in a particular environment.

adipose tissue A living tissue in which many of the cells can store much fat. Fats are chemical substances.

aggression A word confusingly applied, in the behavioural sciences, to many kinds of violence and intolerant activities, even to 'threats' and to defensive territorial warnings such as bird song. Not recommended for serious descriptions of animal or human behaviour.

allele A gene (q.v.) is said to be an allele (or alternative) of another gene when it occupies the same position (locus) on a chromosome (q.v.) but produces a different effect on development. One allele can mutate into another. See also **mutation**.

altruism (1) The primary meaning is a moral one: the principle of acting on behalf of others, regardless of the effect on oneself. (2) In some biological (especially sociobiological) writings, has been given another, quite different meaning: behaviour which lowers the Darwinian fitness (q.v.) of the actor but increases the fitness of another member of the same species; nothing is said or implied about the intentions of the actor. For this category of behaviour, I have elsewhere suggested the term 'bioaltruism' (q.v.). See also **selfishness**.

aposematism Conspicuous appearance accompanied by an unpleasant taste or toxicity. A means by which some animals, especially insects, are protected against predators. See also **mimicry**.

astrology The art of predicting events, especially in human affairs, by study of the planets and stars. According to astrologers, a person's character and fate are influenced by the positions, at the time of birth, of the heavenly bodies (the horoscope). Rigorous tests of astrological beliefs have failed to support them.

average 1. In colloquial speech, typical or usual. 2. Also in everyday speech, the figure derived by counting or measuring objects or events and dividing the total by the number of observations; this, more formally, is the arithmetic mean. (In this book, the mean number of words in a complete line of main text is about 10.) 3. The modal average or mode is the most frequent or commonest score in a series of measures. 4. The median is the middle score in a distribution: that is, the score with as many items below it as above. When the distribution of a series of measurements is symmetrical, as in the Gaussian curve of error, the mean, median and mode coincide. (The reader could plot the distribution of numbers of words to a line in this book, to see whether the distribution is symmetrical.) But some distributions, for instance of incomes, are far from symmetrical: the mean, median and mode are then different. See the illustration on page 152.

axon A long process of a nerve cell which normally conducts impulses away from the cell body.

bacterium (plural: bacteria) Bacteria are microorganisms, nearly always single cells (q.v.), without nuclei (that is, they are prokaryotes (q.v.)). Essential in soil formation and in maintaining the cycles of carbon, nitrogen and other elements. Many are parasites; some cause human disease.

behaviorism The doctrine that the proper subject of scientific psychology is behaviour only. Feelings, intentions, beliefs and other subjective phenomena are disregarded. Even physiological analysis is sometimes rejected.

bioaltruism Behaviour that lowers the 'Darwinian' fitness of the actor but increases the fitness of a member of the same species (q.v.). (In sociobiology, q.v., 'altruism', q.v., is used in this sense.) The behaviour is defined by its effects on others: nothing is said about the intentions of the actor.

biosphere The whole assemblage of living things on earth and the environments that support them.

cell Unit of living matter, usually microscopic, bounded by a very thin plasma membrane and in plants also by a wall of cellulose. Usually contains a nucleus (q.v.); but the smallest organisms (bacteria, q.v.) are cells without nuclei (or prokaryotes, q.v.).

chromosome A body, usually thread-like, consisting mainly of DNA (q.v.) and protein. Chromosomes make up most of the contents of the nucleus (q.v.) of a plant or animal cell (q.v.). In the somatic cells of animals, that is, cells which are not eggs or sperms, and in most plant cells, chromosomes occur in pairs (the diploid condition); the members of a pair are identical in appearance under an ordinary (light) microscope, and are said to be homologous. Eggs and sperms (gametes) have only one chromosome of each pair (the haploid condition). The number of chromosomes varies with the species. Chromosomes are easily seen only during nuclear division.

clone Individuals all of which have been produced, by asexual reproduction, from a single individual. In the absence of mutation (q.v.), all have the same genetical constitution. They are not, however, identical.

control (experiment) In many experimental projects, it is desirable to vary only one condition at a time. Suppose the effect of a dietary substance is being studied: an experimental group, given the substance, will be compared with a control group not given the substance but identical, in every other respect, with the first group.

cryptic appearance Concealment due to resembling a background. A means by which some animals escape the attentions of predators. (Contrast aposematism, q.v.). In some writings, has been absurdly equated with lying.

culture The learned practices, beliefs and attitudes of a human society. These are transmitted from generation to generation, not through the genes but by teaching (q.v.) and other social means. They may be conveyed also between unrelated people. 'Culture', though it has many meanings, represents a concept essential for understanding human existence.

democracy Now a term of almost universal approval, but until recently quite the reverse: from Plato to the nineteenth century, it had derogatory connotations like those of 'communism' today. Originally contained two important principles: opposing despotism and expressing the popular will. Today, it may signify universal suffrage, freedom of speech and assembly and equality before the law (of which only universal suffrage has been achieved anywhere); or it may refer to equivalence of opportunity (still only an aspiration); or some combination of freedom and equality.

dendrite Short, branching projection of a nerve cell. Makes connexions (synapses, q.v.) with the dendrites or axons (q.v.) of other cells.

deoxyribonucleic acid See **DNA**.

DNA Deoxyribonucleic acid. A giant molecule consisting of many nucleotides forming a chain; usually two chains are joined, parallel to each other, and coiled in a helix. Each nucleotide contains one of four bases (thymine, cytosine, adenine or guanine) and a sugar, deoxyribose. Found mainly in the chromosomes (q.v.) of animals and plants, and in the corresponding structures of bacteria (q.v.). The order in which the bases occur in the chain is the 'genetic-code' which, with the intervention of RNA (q.v.), determines the synthesis of proteins. DNA is the material basis of biological inheritance. See also **gene**.

dominant (adj.) (genetics) Properly used, this term is applied to a trait appearing in the phenotype (q.v.) as the result of the presence of a single copy of a gene (q.v.). In the corresponding position (locus) on the homologous chromosome (q.v.), a different form or allele (q.v.) of the gene may be present; this is the heterozygous state and the individual is said to be, in respect of this locus, a heterozygote. When a trait appears only if *two* identical copies of a gene are present (the homozygous state), one on each of two homologous chromosomes, that trait is said to be recessive. In some writings, genes (as well as traits or characteristics) are called dominant or recessive. This usage is not recommended: the effects of genes are usually multiple (see **pleiotropy**); some of the traits they influence may then be dominant but others, recessive.

ecology The science of the relationships of organisms with their environment, including the environment provided by other organisms. An ecologist may study associations of species, as in a rock pool, a patch of soil, a desert, a wood or the human skin; or the changing numbers and density of a population of organisms; or the flow of matter and energy through an association of species; and much else.

ecosystem An association of interacting organisms, together with their non-living environment. Such a system includes producers (green plants) which synthesize organic matter from inorganic; consumers (mainly animals); and decomposers (mainly bacteria and fungi).

epigenesis The appearance of new structures during individual development or ontogeny (q.v.); this entails an interaction of the effects of genes with environmental influences. Opposed to the obsolete doctrine of preformation, according to which the organism is already fully formed in the fertilized egg.

epistasis Modification of the effect of a gene (q.v.) by another, non-allelic gene. An extreme case is when a mutant gene, in one position in a chromosome (q.v.), suppresses the action of a gene in another position.

ethnobotany The study of the relationship between plants and human beings, especially 'indigenous' people who are not technically advanced.

ethology The science of animal behaviour. A major division of biological science, like ecology (q.v.). In some writings used for the study of the behaviour of animals in their natural surroundings.

eugenics Maintaining or improving the quality of a human population by controlled breeding.

eukaryote Of cells (q.v.) or organisms: having a nucleus (q.v.) separated from the rest of the cell (the cytoplasm) by a membrane. Contrast **prokaryote**.

evolution, organic The descent of organisms from very different organisms in the past. Traceable in the fossil remains of organisms. Evolutionary change is still going on; but it is very slow and can be observed in a human lifetime only on a small scale.

exploration In ethology (q.v.), movements of an animal about its living space which are independent of any special need, for instance for food. Such movements represent a tendency to approach strange objects and places and are a means by which an animal learns about its surroundings. In human terms, an exploring animal is displaying curiosity and the desire for stimulation.

fitness (Darwinian fitness) In biology, the fitness of an organism is some measure of its contribution to later generations. There is no necessary connexion with athletic prowess.

gatherer hunters People without agriculture or herds, who live by gathering and hunting food.

gene Originally meant a unit of heredity of which the existence was inferred from breeding experiments. The units are passed on from parent to offspring unaltered. Soon after 1900 (when Mendel's findings were rediscovered), the genes or hereditary factors were shown to be arranged in line in the chromosomes (q.v.) of the cell nucleus (q.v.). Later, the material of heredity was found to be DNA (q.v.). Hence today the word gene usually means a length of DNA. Much chromosomal DNA is, however, 'noncoding': see **transposable element**.

genetical (also **genetic**) Usually means related to the action of genes. May properly refer to *differences* between organisms: some differences are genetically determined. Characteristics or traits are sometimes said to be genetical (or genetic), but this usage is not recommended: *all* traits are influenced by both the genes *and* the environment.

genetic code See **DNA**.

genetics The science of variation and heredity.

genome The whole array of the genetical material or DNA (q.v.) of a species.

genotype The genetical constitution, that is, the whole set of genes (q.v.), of an organism. Contrast the phenotype (q.v.), which is its whole set of actual characteristics.

heritability A measure of the extent to which, in a specified population, genetical differences contribute to the observed phenotypic variation. A high heritability does *not* imply that the trait measured is little affected by environmental change. See also **phenotype**.

homeorhesis Maintenance of a constant sequence of developmental stages.

homeostasis Maintenance of a steady internal state: for instance, blood composition or temperature.

horoscope See **astrology**.

hunter gatherers Alternative name for gatherer hunters (q.v.).

ideology An interpretation of human society combined with proposals for improving it. 'Ideological' is also used as a term of abuse for opinions which differ from the those of the speaker.

induction Reaching a general conclusion from observation of particular instances, especially without bias from previous experience or belief.

instinct 1. In everyday speech, often means the same as intuition or unconscious skill: as when a person is said to avoid a blow 'instinctively' or to understand the attitude of another person 'by instinct'. 2. When an animal performs a complex act without learning how to do it, the act may be called 'instinctive behavior'. In ethology (q.v.), this usage is being given up, for reasons outlined in chapter 4. 3. A third meaning is the same as, or similar to, that of 'drive' or the impulsion to act in certain ways. In some writings expressions like 'hunger drive', 'aggressive drive' and so on still occur. This usage (an echo of vitalism, q.v.) is also being given up: instead, the actual behaviour is described and also, when known, the physiology underlying the behaviour.

kin selection Natural selection (q.v.) is held to favour traits which help both an individual's offspring and also its near relatives.

meme A vague term used for a tradition, belief, technique or practice; it implies, incorrectly, that these are distinct units, analogous to genes (q.v.), subject to a process similar to natural selection (q.v.). See also **culture**.

mimicry Similarity of one species of animal to another, with a protective function. Sometimes, one species is both poisonous or distasteful and conspicuously marked, while the other (the mimic) is only conspicuous (Batesian mimicry). Sometimes both species are toxic and conspicuous (Mullerian mimicry).

mutant See **mutation**.

mutation Sudden change in the DNA (q.v.) usually of a chromosome (q.v.). Mutations can occur in any cell nuclei. Those that occur in a gamete produce mutant DNA which can be passed on to later generations. Many mutations are changes in single genes (q.v.). Mutation is speeded up by radiation and by some poisons. Most mutant genes have an adverse effect on the organism; but some, especially in a changing environment, are advantageous. See **natural selection; nucleus; transposable element**.

natural A word of many meanings. The 'natural environment' of a species means, as a rule, the usual environment of the species and the one to which it is adapted. The expression 'in nature' commonly means in normal or usual conditions. 'Natural' is then opposed to 'artificial'. Problems arise when this distinction is applied to human affairs. Early human beings lived as gatherer hunters (q.v.); this condition may then be said to be natural for the human species. Yet it is also 'natural' (normal, usual) for human beings to change the conditions in which they live (especially by using tools which themselves change with time); people also domesticate and alter the plants and animals on which they depend. Hence nearly all our food and other organic materials are 'artificial' products of human ingenuity.

natural selection The name for processes in nature which result from the existence of genetically determined variation among organisms. As a result of this variation, some organisms contribute more to later generations than do others and are then said to have greater fitness (q.v.). Evolutionary change is held to be largely due to differences of fitness in this sense. But natural selection can also *prevent* change: see **stabilizing selection**.

naturalism (biological) Explaining human social action (and, sometimes, other features) by study of animals: a longstanding custom which frequently leads to absurd errors.

nature Like 'natural' (q.v.), a word of many meanings; not a technical term in biology. Here is a comment by an art historian, Kenneth Clark:

> We are surrounded with things which we have not made and which have a life and structure different from our own: trees, flowers, grasses . . . They have inspired us with curiosity and awe. They have been objects of delight. . . . And we have come to think of them as contributing to an idea which we have called nature.

neuron A nerve cell. Has processes projecting from it, often very many. The cells and their processes conduct impulses among themselves; or from sense organs to a central nervous system; or from the central nervous sytem to muscles or glands. Transfer of impulses from one neuron to another is at junctions called synapses (q.v.). See also **axon**.

nucleus The part of a cell which contains the chromosomes (q.v.).

Present in nearly all the cells of many-celled organisms. See also **eukaryote**.

ontogeny The sequence of changes during individual development. Contrast **phylogeny**.

phenotype All the characteristics of an organism. Contrast genotype (q.v.). Two organisms can have the same genotype, but different phenotypes, because their environments are different. See also **epigenesis**.

pheromone Odorous substance which acts as a social signal.

phrenology Interpretation of human character by the shape of the skull. A longstanding error now obsolete.

phylogeny The sequence of changes during the evolution of an organism or taxon (q.v.).

pleiotropy Multiple actions of a single gene (q.v.). A mutant gene may produce many differences from the normal phenotype (q.v.).

prokaryote Possessing genetical material as filaments of DNA (q.v.) which are not separated in a nucleus (q.v.). Bacteria (q.v.) are prokaryotes. Contrast **eukaryote**.

recapitulation In a narrow sense, means the appearance in embryonic development of features resembling those of an ancestral adult. The embryonic stages of, for example, mammals (including human beings) were once thought to correspond accurately to their evolutionary history. They do not. But the developmental stages of all land vertebrates (which have evolved from fish) do resemble the early stages of a fish. Many other such instances are known.

recessive (genetics) See **dominant** (genetics).

reduction, explanatory The findings of one kind of study can sometimes be partly explained by reduction to those of another. This is a powerful method in all branches of science: for instance, knowledge of the chemistry of DNA (q.v.) helps us to understand the phenomena of heredity. The power of reduction has led to the belief that all biology can be explained by physics and chemistry and to attempts to explain human societies in terms of genes (q.v.) and the past action of natural selection (q.v.). But, if organisms and human beings are to be understood, it is first necessary to make nonreductionist statements about them. These cannot be replaced by biochemical, genetical or other reductionist statements.

ribonucleic acid See **RNA**.

RNA A giant molecule consisting of a chain of many nucleotides each containing one of four bases (uracil, cytosine, adenine or guanine) and a sugar, ribose. Found largely in the ribosomes in the cells (q.v.) of animals and plants, and as messenger RNA and transfer RNA, both of which are concerned in translating the 'coded' information in DNA (q.v.) into protein structure.

scientism The presumption that science is a form of knowledge superior to and superseding all others. See also **naturalism, reduction**.

selfishness 1. In ordinary speech, the principle of acting on one's own behalf, regardless of the interests of others: the opposite of altruism (q.v.) in its primary sense. 2. In some biological writings has a quite different meaning: behaviour which tends to promote the chances of survival of the actor's genes, regardless of the genes of others; nothing is said or implied about the intentions of the actor.

sociobiology The interpretation of animal and human social behaviour by the supposed past action of natural selection. Applied to humanity, it exemplifies both the limitations of explaining by reduction (q.v.) and the errors of naturalism (q.v.).

species The smallest unit of classification or taxon (q.v.) commonly used. All the members of a population assigned to one species, if they reproduce sexually, are usually assumed to be capable of interbreeding. When two kinds of organism are assigned to different species, they are assumed to be incapable of interbreeding or unable to produce fertile offspring.

stabilizing selection If an organism is very different from the typical or normal for its species, it is likely to be less fit, in the biological sense (see **fitness**), than those nearer the norm. But see **natural selection**.

synapse The functional connexion of two nerve cells or neurons (q.v.). 'Connexion' does not signify continuity: a gap always remains between the surface membranes of the two cells.

systematics The classifying of organisms. See also **taxonomy**.

taxon (plural: **taxa**) A unit of biological classification. The principal taxa are, from largest to smallest: kingdom, phylum, class, order, family, genus, species.

taxonomy The principles or theory of biological classification. Also used as a synonym for systematics.

teaching In the present book, the word teaching *in the strict sense* means an activity that alters the behaviour of a member of the same species (the pupil) and tends to be persisted in until the pupil reaches a certain standard of performance or improvement. (In ordinary usage, 'teaching' has many other meanings.) The elaborate teaching of skills is a distinctively human trait.

territory In animal behaviour, a region, occupied by an individual or a group, from which other members of the species are excluded. Often reserved for a region *defended* from others. Territorial behaviour should be distinguished from behaviour which maintains a status system *within* a group of animals. It should also not be muddled up with human ownership of property.

transposable element (jumping gene; transposon) A mobile, virus-like particle, part of a DNA (q.v.) molecule, which influences gene action but, unlike a gene (q.v.), does not code for protein synthesis.

It can alter its position on a chromosome (q.v.) or move outside the cell (q.v.), even to a different species. Transposable elements make at least 10 per cent of the genome (q.v.) of complex organisms. They may be an important source of variation, in addition to mutation (q.v.), and are perhaps significant for our understanding of evolution (q.v.).

ultradarwinism The interpretation of human social action by the supposed influence of natural selection (q.v.). See also **meme, reduction, sociobiology.**

vitalism The long held belief in internal forces, not physical or chemical, as responsible for the life of organisms. Today expressions such as 'instinct' (q.v.) and 'drive' are residues of vitalism.

BIBLIOGRAPHY

Abell G.O. & Singer B. (ed.)1981 *Science and the Paranormal* (New York: Scribner)

Abercrombie M.L.J. 1960 *The Anatomy of Judgment* (London: Hutchinson)

Acton (J.E.E. Dalberg) 1906 *Lectures on Modern History* (London: Macmillan)

Ardrey R. 1967 *The Territorial Imperative* (London: Collins)

Ashby E. 1972 Science and Antiscience. *Sociological Reviews, Monograph* 18

Atkins P.W. 1995 The limitless power of science. In: *Nature's Imagination* (ed. J. Cornwell, Oxford: Oxford University Press) pp. 122–32

Baker R. 1996 *Sperm Wars* (London: Fourth Estate)

Balchin N. 1949 *A Sort of Traitors* (London: Collins)

Barnett J.A. *et al.* 1999 *Yeasts: Characteristics and Identification* (Cambridge: Cambridge University Press)

Barnett S.A. 1988 Enigmatic death due to 'social stress'. *Interdisciplinary Science Reviews* 13, 40–51

Barnett S.A. 1988 *Biology and Freedom* (Cambridge: Cambridge University Press)

Barnett S.A. 1994 Humanity as *Homo docens*: the teaching species. *Interdisciplinary Science Reviews* 19, 166–174

Barnett S.A. 1998 *The Science of Life: From Cells to Survival* (Sydney: Allen & Unwin)

Barnett S.A., Brown V.A. & Caton H. 1983 The theory of biology and the education of biologists. *Studies in Higher Education* 8, 23–32

Barnett S.A. & Dickson R.G. 1989 Wild mice in the cold: some findings on adaptation. *Biological Reviews* 64, 317–40

Basalla G. 1988 *The Evolution of Technology* (Cambridge: Cambridge University Press)

Beer C.G. 1974 Comparative ethology and the evolution of behaviour. In: *Ethology and Psychiatry* (ed. N.F. White, Toronto: University of Toronto Press) pp. 173–181

Boden M.A. 1995 Artificial intelligence and human dignity. In: *Nature's Imagination*, (ed. J. Cornwell, Oxford: Oxford University Press) pp. 148–60

Brannin J.R. 1983 Cognitive factors in children's arithmetic errors. In: *The Acquisition of Symbolic Skills* (ed. D. Rogers & J.A. Sloboda, New York, NY: Plenum) pp. 543–50

Broad W. & Wade N. 1982 *Betrayers of the Truth* (Oxford: Oxford University Press)

Bury J.B. 1913 *A History of Freedom of Thought* (London: Williams & Norgate)

Campbell J. 1989 *The Improbable Machine* (New York: Simon & Schuster)

Carr E.H. 1961 *What Is History?* (London: Macmillan)

Childe V.G. 1947 *History* (London: Cobbett Press)

Clark C.M.H. 1976 *A Discovery of Australia* (Sydney: Australian Broadcasting Commission)

Clark K. 1976 *Landscape into Art* (London: Murray)

Colbert C. 1997 *A Measure of Perfection: Phrenology and the Fine Arts in America* (Chapel Hill, CA: University of North Carolina Press)

Cole G.D.H. 1923 *Social Theory* (London: Methuen)

Cornford F.M. 1957 *From Religion to Philosophy* (New York: Harper)

Cornwell P. 1992 *All that Remains* (London: Little, Brown)

Corson S.A. *et al.* 1975 Pet-facilitated psychotherapy. In: *Pet Animals and Society* (ed. R.S. Anderson, London: Bailliere Tindall) pp. 19–36

Crick F. 1988 *What Mad Pursuit* (New York: Basic)

Cromer A. 1993 *Uncommon Sense* (New York: Oxford University Press)

Crook D.P. 1994 *Darwinism, War and History* (Cambridge: Cambridge University Press)

Crosby A.W. 1997 *The Measure of Reality* (Cambridge: Cambridge University Press)

Dart R.A. 1959 *Adventures with the Missing Link* (New York: Harper)

Darwin C. 1859 *On the Origin of Species by Means of Natural Selection* (London: Murray) [First edition]

Dawkins R. 1976 *The Selfish Gene* (New York: Oxford University Press)

Dawkins R. 1996 *Climbing Mount Improbable* (London: Penguin)

Deacon T.W. 1992 The neural circuitry underlying primate calls and human language. In: *Language Origin* (ed. J. Wind *et al.*, Dordrecht: Kluwer) pp. 121–62

Dehaene S.S. 1997 *The Number Sense* (New York, NY: Oxford University Press)

Dennett D.C. 1995 *Darwin's Dangerous Idea* (New York, NY: Simon & Schuster)

de Solla Price D.J. 1965 The science of science. In: *New Views of the*

Nature of Man (ed. J.R. Platt, Chicago: University of Chicago Press) pp. 47–70

Diamond J. 1997 *Guns, Germs and Steel* (London: Cape)

Dodds E.R. 1956 *The Greeks and the Irrational* (Berkeley: University of California Press)

Doyle A.C. 1893 *The Memoirs of Sherlock Holmes* (London: Newnes)

Dyson F. 1995 The scientist as rebel. In: *Nature's Imagination* (ed. J. Cornwell, Oxford: Oxford University Press) pp. 1–11

Edelman G.M. 1995 Memory and the individual soul. In: *Nature's Imagination* (ed. J. Cornwell, Oxford: Oxford University Press) pp. 200–6

Ekman P. 1985 *Telling Lies* (New York: Norton)

Elton C. 1942 *Voles, Mice and Lemmings* (Oxford: Clarendon)

Elton G.R. 1969 *The Practice of History* (London: Fontana)

Fausto-Sterling A. 1985 *Myths of Gender* (New York: Basic Books)

Francis C.M. *et al.* 1994 Lactation in male fruit bats. *Nature, London* 367, 691–2

Francis D. 1985 *Break In* (London: Pan)

Francis D. 1986 *Bolt* (London: Michael Joseph)

Freud S. 1929 *Civilization and its Discontents* (London: Hogarth)

Freud S. 1949 *An Outline of Psychoanalysis* (London: Hogarth)

Fromm E. 1980 *Greatness and Limitations of Freud's Thought* (London: Cape)

Frye N. 1982 *The Great Code* (New York: Harcourt)

Frye N. 1990 *Myth and Metaphor* (Charlottesville: University Press of Virginia)

Fuller S. 1997 *Science* (Buckingham: Open University Press)

Futuyma D.J. 1979 *Evolutionary Biology* (Sunderland, MA: Sinauer)

Gigerenzer G. *et al.* 1989 *The Empire of Chance* (Cambridge: Cambridge University Press)

Glantz S.A. (ed.) *The Cigarette Papers* (Berkeley: University of California Press)

Gombrich E.H. 1962 *Art and Illusion* (London: Phaidon)

Gould S.J. 1978 *Ever Since Darwin* (London: Deutsch)

Gregory R.L. 1970 *The Intelligent Eye* (London: Weidenfeld & Nicolson)

Griffin D.R. 1958 *Listening in the Dark* (New Haven: Yale University Press)

Haldane J.B.S. 1927 *Possible Worlds* (London: Chatto & Windus)

Haldane J.B.S. 1932 *The Causes of Evolution* (London: Longmans, Green)

Hansel C.E.M. 1966 *ESP: a Scientific Evaluation* (New York: Scribner)

Hayek F.A. 1983 *Knowledge, Evolution and Society* (London: Adam Smith Institute)

Hayek F.A. 1988 *The Fatal Conceit: the Errors of Socialism* (London: Routledge)

Haynes R.D. 1994 *From Faust to Strangelove* (Baltimore: Johns Hopkins University Press)

Hersh R. 1997 *What is Mathematics, Really?* (New York: Oxford University Press)

Hobsbawm E. 1994 *Age of Extremes* (London: Michael Joseph)

Hogben L.T. 1936 *Mathematics for the Million* (London: Allen & Unwin)

Huesmann L.R. *et al.* 1988 The effects of media violence on the development of antisocial behavior. In: *Handbook of Antisocial Behavior* (ed. D. Stoff *et al.*, New York: Wiley) pp. 181–93

Huxley J.S. 1953 *Evolution in Action* (London: Chatto & Windus)

Kevles D.J. & Hood L.E. (ed.) 1992 *The Code of Codes* (Cambridge, MA: Harvard University Press)

Kimble G.A. 1978 *How to Use (and Misuse) Statistics* (Englewood Cliffs, NJ: Prentice-Hall)

Kitcher P. 1985 *Vaulting Ambition* (Cambridge, MA: MIT Press)

Kitcher P. 1997 *The Lives to Come* (New York: Simon & Schuster)

Knight D. 1998 Working in the glare of two cultures. *Interdisciplinary Science Reviews* 23, 156–60

Lagerspetz K.M.J. *et al.* 1988 Is indirect aggression typical of females? *Aggressive Behavior* 14, 403–14

Lévi-Strauss C. 1966 *The Savage Mind* (London: Weidenfeld & Nicolson)

Lewontin R.C. 1979 Sociobiology as an adaptationist program. *Behavioral Science* 24, 5–14

Lewontin R.C. 1991 *Biology as Ideology* (New York: Harper)

Loftus E. & Ketcham K. 1996 *The Myth of Repressed Memory* (New York: St Martin's)

Lorenz K.Z. 1966 *On Aggression* (London: Methuen)

Lovejoy A.O. 1960 *The Great Chain of Being* (New York: Harper)

Mach E. 1959 *Analysis of Sensation* (New York: Dover)

Margulis L. & Schwartz K.V. 1998 *Five Kingdoms* (New York: Freeman)

Medawar P.B. 1967 *The Art of the Soluble* (London: Methuen)

Monod J. 1974 *Chance and Necessity* (Glasgow: Fontana)

Morgan T.H. 1919 *The Physical Basis of Heredity* (Philadelphia: Lippincott)

Mowlana H. *et al.* 1992 *Triumph of the Image: The Media's War in the Persian Gulf* (Boulder: Westview)

Nagel T. 1995 *Other Minds* (New York: Oxford University Press)

Newton-Smith W.H. 1981 *The Rationality of Science* (London: Routledge)

O'Hear A. 1989 *Introduction to the Philosophy of Science* (Oxford: Clarendon)

O'Hear A. 1997 *Beyond Evolution* (Oxford: Clarendon)

Orr H.A. 1996 Dennett's dangerous idea *Evolution*, pp. 50, 467–72

Penrose R. 1995 Must mathematical physics be reductionist? In: *Nature's Imagination* (ed. J. Cornwell, Oxford: Oxford University Press) pp. 12–26

Pinker S. 1994 *The Language Instinct* (New York: Morrow)

Popper K.R. 1959 *The Logic of Scientific Discovery* (London: Hutchinson)

Ramón y Cajal S. 1988 *Recollections of My Life* (New York: Garland)

Raff R.A. 1996 *The Shape of Life* (Chicago: Chicago University Press)

Rauschner F.H. *et al.* 1993 Music and spatial task performance. *Nature,* London 365, 611.

Rawls J. 1971 *A Theory of Justice* (Boston: Harvard University Press)

Richards R.J. 1987 *Darwinism and the Emergence of Evolutionary Theories of Mind and Behavior* (Chicago: Chicago University Press)

Ridley M. 1993 *The Red Queen* (London: Viking)

Rifkin J. 1992 *Biosphere Politics* (San Francisco: HarperCollins)

Ritchie D.G. 1891 *Darwinism and Politics* (London: Swan Sonnenschein)

Rowan A.N. (ed.) 1988 *Animals and People Sharing the World* (Hanover: University Press of New England)

Russell B. 1956 *Portraits from Memory* (London: Allen & Unwin)

Russell B. 1961 *History of Western Philosophy* (London: Allen & Unwin)

Russell B. 1967 *The Impact of Science on Society* (London: Allen & Unwin)

Sagan C. 1996 *The Demon-Haunted World* (London: Headline)

Salomon J.-J. 1985 Science as a commodity: policy changes. In: *Science as a Commodity* (ed. M. Gibbons & B. Wittrock, London: Longmans) pp. 78–98

Scheinfeld A. 1967 *Twins and Supertwins* (Philadelphia: Lippincott)

Shippey T. 1996 Star-gazing. *London Review of Books* 12.xii.96

Singer C. 1928 *From Magic to Science* (London: Benn)

Singer C. 1943 *A Short History of Science to the Nineteenth Century* (Oxford: Clarendon)

Sloboda J.A. 1983 Is music a symbolic skill? In: *The Acquisition of Symbolic Skills* (ed. D. Rogers & J.A. Sloboda; New York: Plenum)

Sorrell T. 1991 *Scientism: Philosophy and the Infatuation with Science* (London: Routledge)

Souden D. 1997 *Stonehenge Revealed* (New York: Facts on File)

Taton R. 1957 *Reason and Chance in Scientific Discovery* (London: Hutchinson)

Taylor P. 1994 *Smoke Ring: The Politics of Tobacco* (London: Bodley Head)

Thomas K. 1973 *Religion and the Decline of Magic* (London: Penguin)

Tinbergen N. 1968 *Curious Naturalists* (New York: Doubleday).

Tobias P.V. 1995 The brain of the first hominids. In: *Origins of the Human Brain* (ed. J.-P. Changeux & J. Chevaillon, Oxford: Clarendon) pp. 61–81

Toynbee A.J. 1946 *A Study of History*, ed. D.C. Somervell (London: Oxford University Press)

Trotter W. 1916 *Instincts of the Herd in Peace and War* (London: Fisher Unwin)

Tupper V. & Williams R. 1986 Unsubstantiated beliefs among beginning psychology students. *Psychological Reports* 58, 383–8

Turney J. 1998 *Frankenstein's Footsteps: Science, Genetics and Popular Culture* (New Haven: Yale University Press)

von Holst D. 1985 Coping behaviour and stress physiology in male tree shrews. *Fortschr.Zool.* 31, 461–70

Weart S.R. 1987 Nuclear fear: a history and an experiment. In: *Scientific*

Controversies (ed. H.T. Engelhardt & A.L. Caplan, Cambridge: Cambridge University Press) pp. 529–50

Weller A. 1998 Communication through body odour. *Nature, London*, 392, 126–7

White T.H. 1954 *The Book of Beasts* (London: Jonathan Cape)

Williams G.C. 1966 *Adaptation and Natural Selection* (Princeton: Princeton University Press)

Wilson E.O. 1975 *Sociobiology* (Cambridge, MA: Harvard University Press)

Wilson E.O. 1979 *On Human Nature* (Cambridge, MA: Harvard University Press)

Wilson E.O. 1992 *The Diversity of Life* (Cambridge, MA: Harvard University Press)

Wittgenstein L. 1922 *Tractatus Logico-Philosophicus* (London: Routledge)

Wrangham R. & Peterson D. 1996 *Demonic Males: Apes and the Origins of Human Violence* (London: Bloomsbury)

Young J.Z. 1964 *A Model of the Brain* (Oxford: Clarendon)

Young R.M. 1970 *Mind, Brain and Adaptation in the Nineteenth Century* (Oxford: Clarendon)

Ziman J. 1994 *Prometheus Bound: Science in a Dynamic Steady State* (Cambridge: Cambridge University Press)

The monthly journal, *Scientific American*, and the quarterly, *Interdisciplinary Science Reviews*, publish articles relevant to the topics in this book.

INDEX